AF669478

WASTE PLASTIC OIL

Dr. Awinash Kumar

Prof. Pradip Lingfa

Dr. Santosh Kumar Dash

Prof. Prafulla Kumar Manoharan

978-81-978285-2-2

ISBN: 978-81-978285-2-2

DEDICATION

This book, "Plastic Waste Oil: A Path to Circularity," is dedicated to the unwavering spirit of innovation and the persistent pursuit of sustainability. It is a tribute to the scientists, engineers, entrepreneurs, and policymakers who are tirelessly working to transform our relationship with plastic waste. We dedicate this work to the countless individuals around the world who are grappling with the devastating consequences of plastic pollution and who strive for a cleaner, healthier planet. May this exploration of plastic waste oil serve as a catalyst for action, inspiring a collective commitment to responsible resource management and a future where waste is not a burden, but a valuable resource. Furthermore, this book is dedicated to future generations who will inherit the consequences of our choices, with the hope that the knowledge and solutions presented here contribute to a world free from the scourge of plastic pollution.

PREFACE

The increasing demand for industrial innovation and advanced technology is due to the global moderation of society. For several decades, 'Energy' has become a demandable research area all over the world. Demand for alternative sources of energy has been long realized. Cutting-edge science and technology have no value if it doesn't serve the purpose. In this regard, several alternative energy resources have been developed and the search is still going on. Amongst various alternative energy resources, waste plastics derived liquid fuel was explored by several researchers to reduce the pressure on fossil fuels and at the same time to control the issues pertinent to non-biodegradable plastic wastes. Plastic wastes are not easily flattened. Waste plastic disposal is a giant challenge for the global environment. The world is now moving toward zero-waste technology following a green and safe ambiance. Incineration and landfill are common and most popular ways of decomposing wastes, but incineration of plastic wastes threat to the next pollution such as emission of CO_2 and soil pollution. As plastics are made of polymers they take several years to demolition due to their non-biodegradable nature. It is time to move forward with waste-to-wealth technology. Energy from waste alteration methods plays a favorable role in zero-waste technology ahead.

The present book explores the experimental production of liquid hydrocarbons by thermo-catalytic degradation of waste high-density polyethylene using kaolin clay, and sodium bentonite as catalysts at lab scale. The types of liquid fuel have been produced by using mutated catalysts. The experimental results show that mutated kaolin clay and sodium bentonite clay have prominent characteristics in the conversion of waste polymers into useful hydrocarbons. Optimal studies explore that the maximum liquid fuel yield is caused by the mixture of acid-treated catalysts. 61.89% of oil yield has been increased with acid-treated catalysts whereas 38.07% of increased oil yield has been calculated with base-treated catalysts. Without using any catalysts thermal degradation of waste high-density polyethylene is recommended for the production of gaseous products. Maximum gas yield has been calculated without using a catalyst.

The liquid fuels produced from the thermo-catalytic waste valorization of high-density polyethylene mainly contain aliphatic compounds. The presence of mutated catalysts greatly supports thermo-catalytic reactions. Finally, the economic analysis of the waste high-density liquid hydrocarbons supports the industrial scale waste plastic oil production, which will boost both rural economy and energy security of the country.

The fuel properties of the produced liquid samples are similar to that of petroleum products. Therefore they can be used as an engine fuel after fractionation or as a feedstock to petroleum refineries.

Dr. Awinash Kumar

CONTENTS

FIGURES

TABLES

SYMBOLS

%	Percentage
°C	Degree Celsius
A°	Angstrom
cm	Centimeter
cm^{-1}	Per centimeter
cm^2	Square centimeter
g/ml	Gram/milliliter
gm	Gram
Gm/mol	Gram/mole
hr	Hour
kg	Kilogram
kV	Kilovolt
kW	Kilowatt
m	Meter
M	Mole
m^2/g	Square meter/gram
mA	Milliampere
ma	Air flow rate
mf	Flow rate of fuel
min	Minute
MJ	Mega Joule
ml	milliliter
ml/min	Milliliter/minute
mm	Milliliter
Nm	Newton Meter
rpm	Revolution per minute
µm	Micrometer
Θ	Diffraction angle

ABBREVIATIONS

ABS	Acrylonitrile
AO	Acid treated thermo-catalytic Oil
AO5	5% AO + 95% diesel fuel (volume)
ASTM	American society for testing materials
ATK	Acid treated Kaolin clay
ATSB	Acid treated Sodium bentonite clay
BO	Base treated thermo-catalytic Oil
BO5	5% BO + 95% diesel fuel (volume)
BP	Brake power
BSEC	Brake specific energy consumption
bTDC	Before top dead centre
BTE	Brake thermal efficiency
BTK	Base treated Kaolin clay
BTSB	Base treated Sodium bentonite clay
C	Carbon
CD1	Commercial Diesel (100 % volume)
CI	Compression ignition
CO	Carbon monoxide
CR	Compression ratio
CV	Calorific value
DAD	Data acquisition device
DI	Direct injection
DTA	Differential Thermal Analysis
EGT	Exhaust gas temperature
FT-IR	Fourier Transform Infrared Spectroscopy
GC-MS	Gas chromatograph and mass Spectrometer
HCl	Hydrochloric acid
HDPE	High-density polyethylene

ID	Inner diameter
K	Kaolin clay
LDPE	Low-density polyethylene
LOI	Loss on ignition
NaOH	Sodium hydroxide
NOx	Oxides of nitrogen
O	Oxygen
OD	Outer diameter
PA	Nylon
PC	Polycarbonate
PET/PETE	Polyethylene Terephthalate
PO	Parent Oil
PO5	5% PO + 95% diesel fuel (volume)
PP	Polypropylene
PS	Polystyrene
PVC	Polyvinyl chloride
SB	Sodium bentonite clay
SEM/EDAX	Scanning Electron Microscope/ Energy Dispersive X-Ray Analysis
TGA	Thermogravimetric Analysis
UHC	Unburned hydrocarbon
WPO	Waste plastic oil
WWT	Waste-to-wealth Technology
XRD	X-Ray Diffraction
ZWT	Zero Waste Technology

Chapter

INTRODUCTION

1.1 General Background

Monetary turn of events, metropolitan development, and creation techniques all assume part in the age of waste plastics. Plastics have turned into an unavoidable part of currer life. These are non-biodegradable polymer compounds essentially made out of carbons hydrogen, and a couple of different components like nitrogen, chlorine, etc. Plastic item are utilized in an assortment of utilizations in our regular routines, including nurseries mulches, covers, sacks, covering, and wiring, as well as bundling, movies, an compartments. Waste plastics are a huge part of civil wastes, as well as modern, fooc and paper wastes. Plastic garbage removal is a major assignment for the norma biological system. A shrewd worldwide populace has surrendered to the ware costs o plastic merchandise. Because of the fast development of non-biodegradable polyme merchandise, strong waste administration and sterilization are currently require measures for living. To keep away from adverse consequences on general wellbeing an the climate it is a difficult assignment of assortment and discarding waste plastics for th state run administrations at civil levels. Burning and landfill are the most widel recognized and well known strategies for garbage removal, however cremation of plasti squanders represents a danger to resulting contamination, like CO_2 outflows and soi contamination. Plastics, which are made of polymers, require quite a long while t deteriorate because of their non-biodegradable nature. Now is the ideal time to take wast to-wealth technology (WWT) to a higher level. Energy from waste plastics procedure will assume a significant part in no waste innovation later on. The expanded interest an creation of high-thickness polyethylene either at the business or modern level has centra issue to the hoarding of its waste. Natural cordial reusing of plastic wastes is the sourc highlight gather or discarding non-biodegradable polymers and can be changed over int an asset of energy. Plastic wastes are quite possibly the most guaranteeing assets fo hydrocarbon. These are not effectively leveled. The world is presently moving toward Zero Waste Technology (ZWT) following a green and safe climate. Towards ideal unmistakable cycles for reusing or reusing plastic wastes is turning into an overall test for the plastic business

1.2 Beginning of the issue

1.2.1 Shortage of non-renewable energy sources

Due to the restricted availability of natural resources of energy, recent increase in the consumption levels of petroleum products is a critical concern for governments. The supply and consumption of petroleum products that have been estimated by different policymakers and professional agencies are not consistent in the present scenario with real facts. At present most of the industries like fuel, transportation, petrochemicals depend upon petroleum feedstocks or crude oil. From this point of view, an alternative path of resource findings is an essential duty for mankind. The utilization and development of alternative energy sources is a favorable act. The alteration of waste polymers into useful hydrocarbon is also a key point followed by the production of transportation like fuel with prominent physicochemical properties.

1.2.2 Solid waste management

Moving towards waste to wealth solid waste management is becoming a milestone in the path of reuse, recovery, and recycling techniques. Combine efforts of law, government policies, ethics, and sustainable strategies are pathfinders of better globalization concerning environmental aspects towards solid waste management activities. Integrated waste management is a reliable way. Soiled, mixed, or off-specification pre-consumer wastes are generated while manufacturing plastic goods from raw materials. Identification, collection, and handling of plastic wastes are done by expert scrap dealers. Uses of secondary polymers are discovered and developed by the dealers depending on outlet markets and applications. Municipal solid wastes cover all types of plastic wastes including agriculture wastes, pharmaceutical wastes, waste furniture, household wares, automobile, electrical and electronic wastes. Household wastes are the major component of municipal solid waste. Municipal solid wastes are heterogeneous, containing a variety of plastic materials including high-density polyethylene (HDPE), low-density polyethylene (LDPE), polystyrene (PS), polyvinylchloride (PVC), polyethylene terephthalate (PET), and others.

1.2.3 Environmental impacts of plastic wastes

The production and disposal of waste plastics have significant environmental consequences. There involves genuine concern about landfilling with waste plastics a non-biodegradable plastic wastes have a long time to break down as well as space requirement. It is obvious that during the production of plastic goods fossil fuels are primary and within the manufacturing the energy resources also require it. It has been evaluated that 4% (per year oil production worldwide) of fossil fuels are used for the production of plastics and 3% to 4% are used within manufacturing. The total environmental impact depends upon the production techniques, applications, and uses of plastic goods. Different types of colorants and harmful chemicals or additives are used within the production of plastic goods. Those are doubtful and not properly checked concerning human lives or surroundings. Waste disposal methods should be as adequate or concerned with as possible as with environmentally friendly. Thermal destruction or incineration releases toxic fumes and harmful chemicals into the environment led to reducing air quality and creating a public health risk. Environmental damage is long-lasting due to plastic wastes. The degradation of marine habitat is increasing due to the continuous release of plastic wastes in oceans. Thermal degradation of waste plastics produces harmful gases.

1.3 Fuels from waste plastics

The valorization of waste plastics into useful products (fuels) is a promising technology for environmental safety. Distant kind of plastic wastes can be employed in the making of valuable petroleum, and use energy resources. The synthesis of liquid hydrocarbon from mixed plastic trash is done in a variety of ways. Table 1.1 shows the list of several plants in India working on waste plastic conversion. Several individual or self financed small scale plants are also running. In fact waste valorizations of polymers are a new approach for the green surroundings. The perspective of waste plastic conversion is about to the separation of different types of compounds (useful hydrocarbons).

Table 1.1 Manufacturing plants of fuels from waste plastics in India

Plants	Products
Waste Plastic to Fuel Oil Plant, Trimurti Production, Indore, Madhya Pradesh, India.	Fuel oil, black carbon powder
Baswar plant, prayagraj, India.	Diesel oil
Waste plastic to Diesel plant, IIP Dehradun, India.	Liquid Fuel
Plastic to Diesel conversion plant, Mathura, India	Petroleum hydrocarbon
Waste to Energy plant, Bhubaneswar, India.	Liquid petroleum products
Rudra Environmental Solutions, Pune, India.	Liquid hydrocarbons
Ventana, Ramky plastic-to-fuel plant, Hyderabad, India.	Liquid Fuel
M.K. Aromatics Limited, Chennai, India.	Petroleum products

1.4 Advantages of production of waste HDPE plastic oil

The rationale behind choosing waste high-density polyethylene for liquid fuel production is due to the following advantages:

- Easily availability of waste high-density polyethylene.
- Does not require land.
- Creates employment, in rural as well as urban areas.
- Reduce dependency on oil imports.
- Reuse, recycling, and recovery strategy for environment safety.
- Foreign exchange savings.
- Regional and national development.
- Provides energy security in the country.
- Green, eco-friendly technology and recovery of energy sources from plastic wastes.
- Cost-effective in comparison with the available fuel resources.

1.5 Research and development on the recycling of plastic wastes in India

Owing to the paucity of fuel resources and limited stocks of transportation fuels, plastic wastes conversion is a good initiative in the way of a green environment. Several development works have been carried out concerning useful hydrocarbon production from plastic wastes. Waste plastic oil has been attracting a large section of researchers due to its multidisciplinary scope. Researchers belong to mechanical sciences, chemical sciences, and environmental sciences exploring and developing the science an technology of waste plastic oil production and its utilization as a transportation fuel t reduce environmental pollution. Indian Institute of Petroleum (IIP) team is working on project to convert plastic wastes into useful petroleum products. GAIL (India) ha sponsored the project for technical support. A patent for pyrolysis of polymer waste ha been developed by the CSIR-Central Mechanical Engineering Research Institute i Durgapur, West Bengal. Several investigation reports from the National Institute c Technology, Rourkela, India (department of chemical engineering) have been publishe on the manufacturing and use of waste plastic fuels (Panda AK 2011). In North-Ea Region at Assam, there is an organization set up by the government of India called Plasti Waste Management Centre (PWMC) which is conducting different short-term courses fc training purposes of conversion methodology of plastic wastes into useful product Laboratory for Advanced Research in Polymeric Materials (LARPM), Bhubaneswar ha developed eco-friendly technology for the categorization of plastic wastes as well a recycling for commercial usage.

1.6 Economics of production of waste plastic oil from HDPE waste

A transportation fuel price is consecutively increasing since the last decade. Liqui hydrocarbons obtained from the different recycling processes of waste plastics have bee recognized by a prominent alternate transportation fuel. It has the potential to be used as supplementary fuel due to its ability to combat the threats posed by the "twin crisis, namely the reduction of hydrocarbon deposits and environmental damage. Howeve waste plastic fuel price is the principal obstacle that affects the commercialization of th product in the market. To tackle this issue, lower-cost feedstocks such HDPE, LDPE PET, and PVC have been exploited which proves to be promising resources for th production of WPO. The type of production methods and catalysts used for the synthesi also plays an important role that contributes to the economics of waste plastic oi preparation. Recently, heterogeneous catalysts are gaining wide attention for their variou advantages compared to their counterpart homogeneous catalysts due to Lower cost easily available, can be derived from natural and waste resources.

Several countries have adopted policies to promote plastic waste management. The Indian government is also taking initiatives under Swachh Bharat Abhiyan for the no plastics campaign. From time to time governments of different countries are companioning for 'say no to plastic'. 97% of plastic bottles in Norway are recycled through the Norway government bottle scheme (climateaction .org, 2018). The ministry supports R&D, pilot plants, demonstration projects leading to the commercial development of waste plastic recycling. Apart from direct economic benefits, secondary economic benefits i.e., rural employment, positive effect on greenhouse gas emission, and improving the energy security of the country will play a vital role in the longer run.

1.6.1 Plastic Recycling -Possible Business Options

The business strategy of the useful hydrocarbons produced from plastic wastes is a good option in the fuel market. It has been found by different news sources that the investment of 25 lakhs is needed to establish a small plant to recycle plastic wastes (PET) into useful hydrocarbons. It is an environment-protecting, low-budget, and emerging zero-waste business opportunity for entrepreneurs, governments, and local bodies. However different types of opportunities and threats have been discussed for the betterment of the market risks as reported in Table 1.2, Table1.3, and Table 1.4.

1.7 HDPE as a promising feedstock for fuel production

High-density polyethylene (HDPE) is a thermoplastic resin derived from petroleum products. In contexts of energy and raw materials, 1 kg of high-density polyethylene requires 1.75 kg of petroleum. In comparison to Low-density polyethylene (LDPE), HDPE has fewer branches with high strength properties and a wide range of applications. HDPE was quickly adopted by Karl Ziegler, a German researcher. He used a catalyst to create HDPE at atmospheric pressure. Aluminum triethyl (a metal alkyl) and titanium tetrachloride were mixed together (organometallic). In the presence of a neutral solvent, ethylene had been added, polyethylene of very little branching, and high molecular was produced. A more crystalline structure had been found. The polymerization reaction took place at 100 °C and atmospheric pressure (Singh B *et al.* 2008). General household plastic goods are HDPE (bucket, jug, bottles, etc.). Due to easy availability of waste

HDPE the conversion cost may be prominent and reliable for the marketing of extracte liquid fuel from it. The promising concern is about to the cost effective feedstock for fue production. HDPE wastes are easily available.

Table 1.2 Production of Furnace fuel from plastic wastes

Opportunity	Threats
• Wider applicability.	• Low-profit margins.
• Lower production cost.	• More stringent pollution control requirements at the point of use.
• Less stringent quality norms.	• Quality is affected by climatic conditions.
• Simple storage regulations and standards.	• Extensive use of innovations.
• Use in large quantities.	
• The industry got spreading around across country.	

Table 1.3 Liquid fuel production (diesel)

Opportunity	Threats
• Wider and rapidly expanding supply chain	• Stricter quality specifications.
• Enhanced profitability	• Non-acceptance by the government and end users.
• Good existing distribution network.	• Part of essential commodity act.
	• Conflict with new fuel prices, which are highly volatile and largely supported by the Indian government.

Table 1.4 Captive plant with converted fuel

Opportunity	Threats
• More nimble tasks at the micro scale.	• Competitions with other renewable energy options.
• Lower investments.	• Dependent on one recycling supply chain.
• Easy to control.	• Discomfort.

Table 1.5 lists the resins that are suitable for thermal degradation (Beyene HD, 2014). Because of their lower sulphur content, liquid hydrocarbon ingredients may be advantageous for renewable fuels. Consea et al. reported deformation of HDPE and the initiation of gaseous products at various temperatures (500 °C to 900 °C). It was determined that at high temperatures, the total gas yield decreased. At high residence times, benzene, methane, and toluene formed well, but propane, butane, propylene, ethylene, ethane, pentane, and butylenes formed to varying degrees depending on temperature rise (Consea JA *et al.* 1994). Walendziewski *et al.* investigated polyethylene thermal cracking at temperatures ranging from 370 °C to 450 °C (Walendziewski *et al.* 2001). It was found that increasing temperature is caused for the fast production rate of gaseous products and liquid products but less amount of residue was found at high-temperature ranges with slow production of gas. Table 1.6 shows the composition of gaseous products found at 400 °C by thermal cracking of polyethylene polymer (Marcilla A *et al.* 2001).

1.8 Research motivation

Waste plastic oil is burning fuel, safe for the environment, produces lesser air pollution, and contains elements like petroleum. Both costs of liquid hydrocarbon production from waste high-density polyethylene and separate feedstock availability are major concerns for which the practical applicability has not reached its potential in many countries. Waste high-density polyethylene is the easily available feedstock for the production of liquid hydrocarbons. Cost issues can be further alleviated by using heterogeneous catalysts for fuel synthesis. Separate collection strategy of waste high-density polyethylene from municipal solid waste tends to lead the easily available feedstock. Minor modification of waste high-density polyethylene and catalyst ratio can be advantageous for the further production of liquid hydrocarbons. Therefore more research is necessary to develop waste HDPE oil (fuel/liquid hydrocarbons) from locally available low cost feedstock and investigate its diesel engine application potential. Therefore more research is necessary to produce low-cost waste plastic oil from waste high-density polyethylene following suitable catalysts and investigate its application potential as transportation fuel like petroleum products.

Table 1.5 Possibility of product outcomes by thermal degradation of plastic wastes

Products	Plastics	Suitable feedstock for liquid fuel
Liquid Hydrocarbons	High-density polyethylene (HDPE)	Yes
	Low-density polyethylene (LDPE)	Yes
	Polypropylene (PP)	Yes
	Polystyrene (PS)	Yes
	Polymethylmethacrylate (PMMA)	Yes
	Acrylonitrile-butadiene styrene copolymer (ABS)	No
Solid products	Polypropylene terephthalate (PET)	No
Carbonous	Polyurethane (PUR)	No
	Phenol resin (PF)	No
Hydrogen chloride and carbonous products	Polyvinyl chloride (PVC)	No
	Polyvinylidene chloride (PVDC)	No

Table 1.6 Gaseous elements at 400 °C by thermal degradation of polymer

Components	Catalytic degradation	Thermal degradation	Hydro Cracking
Methane	12.4	22.7	21.1
C3	30.4	26.6	23.7
C4	20.3	11.0	20.7
C5	5.6	6.9	7.3
C6	3.3	2.1	3.8
Ethane	20.4	27.4	21.2
Ethylene	2.3	1.4	0.1

Chapter 2

RESEARCH SYNTHESIS

2.1 Introduction

This section investigates a far reaching writing audit on the change of waste plastic into useful hydrocarbons. Starting from feedstock selection to waste plastic oil production and their characterization analysis has been described. All the related information pertinent to the oil production from waste plastics has been summarized and the existing research gap has also been identified. This chapter leads to a set of well-defined objectives and routes to achieve those. The first section deals with plastics and their sources. The second section overviews the recycling techniques used for the conversion of waste plastics. The third section presents the significance of the catalysts. The fourth section explores the mechanism and kinetic characteristics of thermal cracking/depolymerization of waste high-density polyethylene and performance evaluation of waste plastic fuels powered diesel engine. Finally, the research gap and objectives of the present investigations have been covered.

2.2 Plastics

Plastic is one of the handiest, generally dependable, modest, and far and wide materials made by humanity. It pervades all aspects of human life. Because of modern urban expansion and revenue increases, most developing nations are consuming an increasing amount of plastic goods. Plastic products are composed 20 times more present than they were five decades ago. Daily need items like carrying bags, glasses, plastic bottles, and milk and food containers are generally 78 wt% of household plastic wastes. It may be one of the types of HDPE, LDPE, PP, and PS (Achilias DS *et al.* 2004). Plastics are made of petro materials and non-biodegradable. There are two common types of plastic wastes are available such as thermoplastics and thermosetting plastics. In 1988 Society of the Plastics Industry (SPI) made an identification system to associate different kind of polymers. Proper SPI codes are marked by the companies usually at the bottom of the polymer goods. It relates to a guideline of the different polymer goods associated with each code number. As per the available resources scientists have designated the plasti goods as follows (Sarker M *et al.* 2012).

(A) Polyethylene Terephthalate: In daily life uses mineral water, fizzy drink, and bee bottles, some shampoos, and mouthwash bottles are made of PETE or PET. Pre-prepare food trays and roasting bags are also a type of PETE. It has good moisture barrie properties and high heat resistance. Alteration of PETE is accepted. Output as high quality products can be compared with origin polymers (Al-Salem SM *et al.* 2009). It i clear, hard, and well with microwave transparency solvent resistant. It is identified b digit 1 in a triangle shape.

(B) High-density Polyethylene: Bottles of detergent, milk, non-carbonate drinks, bleach fabric conditioner, toys, buckets, and pipes, are generally a kind of HDPE. It ha excellent chemical resistance and moisture barrier properties. It has a soft waxy surface and good permeability to gas. It is identified by digit 2 in a triangle shape.

(C) Polyvinyl Chloride: Credit cards, some synthetic leather products, pipes and fittings carpet backing, and floor covering are a group of PVC. Some types of window and doo frames are also made of Polyvinyl chloride. It has excellent transparency with goo chemical resistance and low gas permeability. Due to its stable electrical characteristic and acceptable weathering ability, it is put-to-use in wire and cable sheathing. It i identified by digit 3 in a triangle shape.

(D) Low-density Polyethylene: Due to its moisture barrier behavior and goo transparency nature, it is commonly used to make irrigation pipes. It is tough, flexible a well as good stable electrical properties so used in making thick shopping bags, an electric wires. It has a soft and waxy surface. Films, bubble wrap, flexible bottles, an some refuse sacks are medley of LDPE. It has a low melting point. It is identified by digit 4 in a triangle shape.

(E) Polypropylene: Most heavy-duty bags, crates, plat pots, bottles top, biscuit wrappers, drinking straws, ketchup, and syrup bottles are set of PP. It has a flexible and waxy surface as well as excellent chemical resistance characteristics. PP is less stable than LDPE and HDPE. It is due to the mechanism of degradation of polymers and CC bonds

(Miskolczi N *et al.* 2004). It is translucent, hard, and strong. It has a high melting point. It is identified by digit 5 in a triangle shape.

(F) Polystyrene: PS has a high clarity and is influenced by fats and solvents. It may be rigid or foamed. It is hard and brittle. It is generally used to make egg boxes, disposable trays, video cases, disposable cups, hangers, and brittle toys. It is identified by digit 6 in a triangle shape.

(G) There are some special types of polymers used in engineering sectors. These are labeled with numbers ranging from 7 to 19 and triangles. Acrylonitrile butadiene styrene (ABS), Polycarbonate (PC), Nylon (PA), and Layered or multi-material mixed polymers are some kinds of it. Digits from 7 to 19 are assigned to the polymer goods as per the applications in the different sectors.

2.2.1 Intent and Sourcing of Plastic Wastes

Inventory management is essential for storing massive amounts of plastic in order for the reusing system to function properly. Plastic waste's true capacity ought to be lined up with advertising income. Dismissed or waste plastic materials from the assembling of plastic products have more noteworthy reusing or fuel asset properties. On account of zero-squander innovation, horticultural squanders like plastic sacks, plastic holders, hosepipes, and sheets can be used as fuel assets.

Craftsmen stores and supermarkets sell a variety of plastics, most of which are polyethylene (PE). Municipal garbage, on the other hand, is an important source of plastic waste. Cities' local scrap yards have a considerable amount of damaged plastic furniture that has been collected by scrap buyers as they move about the area. Drug or clinical waste, modern junk, development trash, and electronic waste (EW) are generally normal wellsprings of plastic waste (ACCR 2004). However, municipal arrangement for the waste polymer collection is the best suitable option.

2.2.2 Pre-use Plastic (Production Scrap)

Pre-utilize plastic trash is in all likelihood made of plastic. These don't meet the determinations for their expected utilization or contain off-cuts from gathering or establishment. These materials might be helpful for elective applications and can possibly be reused, despite the fact that they are not fitting for the expected purposes. This form of plastic trash is the primary source of processable plastics from plastic product makers. It is usually worth more than post-consumer plastic garbage. In most cases, it only requires minor processing before being used in a new product (Aguado J *et al.* 2007, Garforth A. *et al.* 2004, Kyrikou I *et al.* 2007).

2.2.3 Post-use Plastic

Recycling categories of Post-use plastics:

- Water Bottles, pots, tubes, and trays.
- Plastic films.
- Pipes and Rigid plastics.
- An expanded polystyrene (EPS).
- Flexible plastics, such as strapping and cable sheathing (LDPE).

2.2.4 Selection of Waste Plastics: A Promising Resource of Fuel Production

Many investigations have checked out at the requirement for flammable and non-dangerous feedstock in the alteration of waste into fuel. The method for changing plastic over to not entirely set in stone by the sort of plastic chose. For the identification of various kinds of plastics, a division cycle is likewise completed. For appropriate identification, tests with water, burning, odor after burning and scratch are performed. Plastics of all types are not recyclable. Thermo sets, also known as thermosetting, can only be heated once before taking shape (Swift G *et al.* 2004). Repeated heat treatments are incompatible with these polymers. Thermoplastics (PE, PVC, and PP) are an excellent fuel source. Thermoplastics also include polytetrafluoroethylene (PTFE). Plastics that are recycled include the following:

- High-density polyethylene (HDPE)
- Low-density polyethylene (LDPE)
- Polystyrene (PS)
- Polypropylene (PP)

- Polyvinyl chloride (PVC)

2.3 Plastic to Fuel Strategies: Reusing of Waste Polymers

Plastic trash reusing is a significant stage toward a more manageable climate. As a general rule, it was found that there are four procedures to reuse plastics:

2.3.1 Primary Recycling

Plastic trash is transformed into goods with performance equivalent to that of original virgin plastic products. It's quite costly. It's a type of mechanical reprocessing in which the finished product resembles the virgin. This procedure may work well with scraps that aren't completely clean. This isn't the right way to go about things.

2.3.2 Secondary Recycling

The quality standards for modifying plastic wastes are lower than for the original material. This procedure is carried out to create wealth out of garbage. In comparison to the original, the procedure uses different products.

2.3.3 Tertiary Recycling

In the realm of polymer to gas conversion, it is the most often utilized approach. Depolymerization (thermal degradation) is used to create chemicals, fuels, and other related products from plastic trash. The original value of the raw material is lost. Thermal cracking can be used to achieve a high level of waste pollutants in this process.

2.3.4 Quaternary Recycling

In the energy recovery process from plastic trash, incineration plays a critical role. The age of energy in this interaction brings about burning. This technique is more advantageous for managing air contamination than it is for managing strong waste issues. This procedure is avoided by using a solid waste management system. Landfilling accounts for 20% by weight and 10% by volume of virgin trash (Patni N *et al.* 2013).

2.3.5 Pyrolysis

Pyrolysis is an endothermic reaction that can occur with or without the need for catalyst. Before pyrolysis, conversion parameters are required for resin selection. 350 600 °C is the operational temperature range. Furthermore, according to a State Researc Inc. (NSR) analysis report, pyrolysis without any catalyst is the most suited and feasibl method for converting polymer into fuel resources (Silverio FO *et al.* 2008). Impuritie like chlorine, sulphur, and residues are present in the fluid fuel produced by simpl pyrolysis, which can be reduced by employing a catalyst. Pyrolysis completely deliver carbon, condensable fluids (tar), and non-condensable gases. Pyrolysis is carried ou utilizing different kinds of plastic wastes using an impetus, making them proper for th transformation of plastic waste into energy assets. The mechanism of thermal degradatio is made up of four types of kinetics in pyrolysis. Figure 2.1 depicts a straightforwar waste plastic conversion technology.

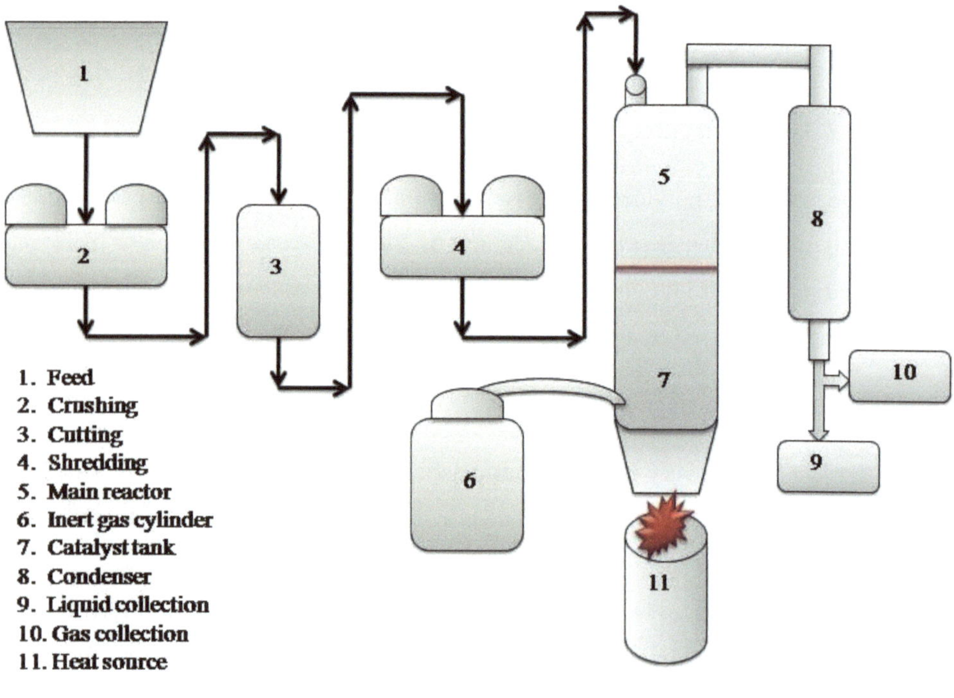

Figure 2.1 General set-up for production of waste plastic oil

The following methods have been proposed: (1) end-chain scission, (2) fragmentation, (3) side-chain removal, and (4) cross-linking. The cross-linking mechanism is preferred in thermosetting polymers. It is dependent on the polymer kinds (Singhad RP *et al.* 2011). Sekine *et al.* and Panda *et al.* proposed using a Fe/activated carbon catalyst to catalyze the degradation of PP by initiating random C-C bond breakage, propagation, and termination. Because the chemical structures of polyethylene (PE) plastics such as HDPE, LDPE, and PP are crossed chain hydrocarbons, pyrolysis is more complex (Sekine Y *et al.* 2003, Panda AK *et al.* 2010a, Di Blasi C 2000).

Pyrolysis has been utilized to improve the liquid by-product derived from it using catalysts such as Red Mud, FCC, HZSM-5, Y-zeolite, Fe_2O_3, Al_2O_3, $Ca(OH)_2$, and natural zeolite. The role of the catalyst in heat degradation is responsible for a higher proportion. In addition, thermo-catalytic degradation uses less energy than thermal degradation (Beyene HD 2014). In the arranged pyrolysis of plastic squanders, MCM-41 and ZSM-5 zeolite impetuses are more helpful for gas. Because of their low sulfur focus, fluid hydrocarbon items might be invaluable for making liquid fuel. Thermal pyrolysis can carry a wide spectrum of hydrocarbons from C_5 to C_{80}. High temperatures are used to generate products of minimal commercial value, such as I-olefin and n-paraffin. Waxes are used to obtain heavy oils. The main disadvantage of this method is that it produces hydrocarbons at concerning temperatures 500 °C to 900 °C. Thermal pyrolysis can also be done in two steps (Htet MT 2010). Thermal cracking is a new method for transforming waste polymers into an energy source (McCaffrey WC *et al.* 1995, Uddin MA *et al.* 1997, Garforth AA *et al.* 1997, Grieken RV *et al.* 2001, Hwang G *et al.* 2001, Navarro R *et al.* 2003, Gao Z *et al.* 2003, Wallis MD *et al.* 2007, Cho MH *et al.* 2010).

Following advantages of pyrolysis process: (Singhad RP *et al.* 2011)

- Low capital expense is gotten.
- First and simple cycle to recuperating energy assets from squander.
- It upholds zero-waste innovation.
- It decreases the volume of waste in the scope of 50-90%.
- Simple to move and store.
- A positive technique to create income from squander.

- For lighter hydrocarbons, high reaction temperature is required; it is the significa downside of thermal cracking (Mohanraj C *et al.* 2017).

2.3.6 Gasification: A Thermo chemical conversion

Gasification is a first-order reaction that happens at temperatures between 600 °C an 800 °C, usually in a low-oxygen atmosphere. In practice, gasification can be eith endothermic or exothermic. Syngas, otherwise called producer gas, is a by-product c gasification (Brems A *et al.* 2013). With a determined fuel briquette, 1.76 MJ/m^3 c energy and cold gas productivity of around 66% were accomplished from 55% of plasti wastes utilizing the gasification technique. The fractional oxidation of natural matter fc the age of syngas is known as gasification. The first stage involves the partial burning c biomass to produce gas and char. In the presence of a gasification agent, heterogeneou reactions occur (Beyene HD 2014). Gasification systems are made up of three primar components:

- The gas cleaning plans; eliminates unwanted mixtures from the ignitable gas.
- The gasification framework; assists with creating flammable gas.
- The energy recuperation plan; recuperates the ideal energy from the cycle.

2.3.7 Thermal Cracking

Thermal cracking is a depolymerization/degradation of polymers that takes place a concerning temperatures 450 °C to 800 °C. The products of thermal cracking are in th form of condensable hydrocarbon liquid, non-condensable gas, and carbonized char Thermal cracking converts larger hydrocarbons into smaller hydrocarbons. Stear cracking is also a type of thermal cracking which uses heat energy generated from th steam, called pyrolysis in the absence of oxygen. It is a method for the production o alkenes caused by the latent heat of energy of the steam. Modern thermal cracking use high pressure and high temperature and hemolytic fission takes place within the carbo bonds. This technique is generally performed in the production of needle coke and burne fuels.

The chain scission of polymers to shape essential petrochemical compounds at different temperatures, response periods, and tensions, and in the presence or nonappearance of receptive gases and impetuses is known as thermal cracking. Cracking the bonds requires heat, which brings about the development of an unstable piece and a carbonized singe, which can then be isolated into non-condensable gases and condensable hydrocarbon oil. Chemical recycling, thermolysis, thermo-catalytic cracking, and gasification of solid plastic wastes have all been utilized to recuperate different hydrocarbon parts in the fuel range (Lee KH 2006, Zhang GH *et al.* 2006 Al-Salem SM *et al.* 2009, Abbas-Abadi MS *et al.* 2012). The peculiarity of heavy hydrocarbons after thermal degradation can be explained as follows:

- Production of gas and coke.
- Slow reaction in comparison with catalytic reactions.
- At high temperatures, some olefins are produced.

2.3.8 Catalytic Cracking

A catalytic cracking method uses a catalyst or a mixture of catalysts in a liquid phase or vapour phase contact, as the name suggests. The catalyst comes into contact with dissolved polymers in the fluid stage, and in the fume stage, the raw polymer or plastic is degraded into hydrocarbon vapours, which then come into contact with the catalyst. There are three sub-processes are involved in the methods as follows:

- Reaction.
- Regeneration.
- Fractionation.

Hydro-cracking (special case of catalytic cracking) is done at high temperature and pressure (Karagoz S *et al.* 2003, Lin YH *et al.* 2004). Different feeds are used to produce gasoline by hydro-cracking. Typical feeds include polystyrene, polyvinylchloride, scrap rubbers, solid wastes, etc. Relative rates of reactions and the type of feedstock are the prominent factors in this method.

Fluid catalytic cracking is a different kind of catalytic cracking. It's done at a hig temperature and a moderate pressure. The powdered and hot catalysts are in contact wit the feedstock. The shorter-chain compounds were made with major catalysts from th long-chain ones.

2.4 Significance of catalysts used in waste HDPE valorization

ZSM-5 zeolite catalyst was used by Sharratt PN *et al.* in thermal degradation of HDP and it was found that oil yield was decreased and gas yield was increased (290 °C to 43 °C) (Sharratt PN *et al.* 1997).

Bagri R *et al.* investigated thermal decomposition of plastic waste (polyethylene wit zeolite and concerning temperatures 400 °C to 600 °C). It was done for the catalyti upgrading of the gaseous products. It had been concluded that coke and liquid yiel decreased and the gas yield increased due to high catalyst bed temperature (Bagri R *et a* 2002). Another investigation was done by Venuto PB *et al.* and found that (concernin temperature 480 °C to 590 °C) a higher amount of alkene gaseous products were obtaine and less formation of coke was reported with zeolite catalyst (Venuto PB *et al.* 1979).

Silica, silica-alumina, zeolites, and alumina were concentrated in the polyme debasement inside the temperature goes from 200 °C to 600 °C by Beltrame PL *et al.* I was concluded that more gaseous products were obtained with high active silica-alumin at 400 °C within the degradation (Beltrame PL *et al.* 1989).

Miskolzi N *et al.* reported the significance of catalysts in the degradation of waste HDPE (clinoptilolite, HZSM-5, and equilibrium FCC). The catalytic degradation process wa done at 430 °C for 45 minutes in each circumstance. Low sulfur and nitrogen conten liquid hydrocarbons and residue were obtained. It was concluded that FCC an clinoptilolite catalysts augmented liquid yield, whereas HZSM-5 boosted both the yield of liquid hydrocarbons and gaseous products (Miskolzi N *et al.* 2008).

Lee KH *et al.* investigated the yields percentage of gaseous products, liqui hydrocarbons, and residue separately for thermal degradation and catalytic degradation.

Table 2.1 Thermal degradation and catalytic degradation of waste HDPE at 430 °C

Types of degradation	Gas %	Decrement after using FCC catalyst	Liquid %	Increment after using FCC catalyst	Residue %	Decrement after using FCC catalyst
Thermal degradation	20.0	3	75.5	5.56	4.5	80
Catalytic degradation	19.4		79.7		0.9	

The reaction temperature was 430 °C. It was reported that catalytic degradation was better than thermal degradation to obtain more yield % of liquid hydrocarbons and less yield % of residue. 5.56% of increment was calculated for liquid yield % after the catalytic degradation shown in Table 2.1. 80% and 3% of decrement were observed for residue yield % and gas yield % respectively.

Without any change in the increase of reaction time paraffin, olefin, and aromatic compounds were obtained in case of thermal degradation whereas, in the case of catalytic degradation gasoline range of light olefin was mainly found. The atomic weight goes from 80 to 400 and less than 200 fluid items were gotten individually for warm and reactant debasement of waste HDPE (Lee KH *et al.* 2003b).

Seo YH *et al.* examined the fluid items got from thermo-reactant debasement of waste HDPE By utilizing ZSM-5, zeolite-Y, mordenite, and nebulous silica-alumina impetuses with GC-MS. A high measure of coke was found in the wake of utilizing mordenite among all. Weighty hydrocarbons were handily broken with nebulous Silica-alumina and created lighter olefins. ZSM-5 showed the best outcome in the arrangement of aromatics and little vaporous hydrocarbons. It was inferred that zeolite and silica-alumina increased the olefins in the fluid item. Table 2.2 shows the different items got at 450 °C with various impetuses. Table 2.3 shows the PONA (Paraffin, Olefin, Naphtane, and

Aromatic) distribution in liquid products obtained at 450 °C with different catalysts (Se YH *et al.* 2003).

Different types of zeolites were studied within (380 °C to 410 °C) cracking of HDPE. had been reported that the pore construction of zeolites assumes a significant part i controlling the item dispersion during polymer breaking. It was observed that the larg size of zeolite enhanced the yield of liquid products (Park JW *et al.* 2002).

Manos G *et al.* researched the thermo-synergist corruption of HDPE with various zeolite and created the C_3 and C_{15} scope of hydrocarbons. It was observed that medium-por zeolites created lighter hydrocarbons in correlation with huge pore zeolites (Manos G *et al.* 2000).

Hernandez MR *et al.* researched the thermo-catalytic degradation (in the absence of oxygen) of high-density polyethylene over HZSM-5 impetus (model device was utilized Two milliseconds of unstable home time was thought of. A fluidized bed reactor wa utilized to get propane, isobutene, and isopentane. C_1-C_3 scope of hydrocarbons fo certain heavier mixtures like n-heptanes, n-hexane, butadiene, and benzene expande with home time with HZSM-5 impetus (prototype apparatus was used). Two millisecond of volatile residence time was considered. A fluidized bed reactor was used to obtai propane, isobutene, and isopentane. C_1-C_3 range of hydrocarbons with some heavie hydrocarbons such as n-heptanes, n-hexane, butadiene, and benzene increased wit residence time with HZSM-5 catalyst. It was reported that the better contact of polyme and catalyst favored the production of light hydrocarbons and gas fractions, wherea reducing gas fractions for low contact (Hernandez MR *et al.* 2006).

Concerning temperature variation, 290 °C to 430 °C catalytic pyrolysis of HDPE wa done by Garforth AA *et al.* in a fluidized bed reactor. Centered on feed the yield o volatile products had been reported over the different catalysts as HZSM-5 > HUSY HMOR > SAHA. The acidic zeolite catalysts HMOR, HUSY, and HZSM-5 showe better results in degradation in comparison with less acidic catalyst SAHA. Bette selectivity of hydrocarbons in the range of C_3-C_5 was found with HZSM-5 and HMOR

The lowest conversion was found with the SAHA catalyst and produced the C_3-C_8 range of hydrocarbons.

Thus the production of potentially valuable hydrocarbons catalysts had the prominent effect to reduce the reaction time and provided better selectivity of products as well as increased volatile products (Garforth AA *et al.* 1998).

Two distinct impetuses to HDPE feed proportion 2:1 and 6:1 in a fluidized bed reactor separately under nitrogen and environment had been concentrated by Ali S *et al.* C_3-C_8 scope of hydrocarbons were gotten with ZSM-5, US-Y, ASA, New FCC, and balance FCC impetuses (Ali S *et al.* 2002).

Table 2.2 The distribution of products obtained at 450 °C with different catalysts

Type of pyrolysis	Yield of Products		
	Liquid (Weight %)	**Coke (Weight %)**	**Gas (Weight %)**
Thermal Cracking	84.00	3.00	13.00
Zeolite (Powder)	71.50	1.50	27.00
Si-Al (Powder)	78.00	1.00	21.00
ZSM-5 (Powder)	35.00	1.50	63.50
Alumina(Powder)	82.00	2.10	15.90
Zeolite (Pellet)	81.00	1.50	17.50
Mordenite(Pellet)	78.50	18.50	3.00

With nanocrystalline HZSM-5 catalyst depolymerization of plastic (HDPE) was investigated by Mastral JF *et al.* within the temperature ranges from 350 °C to 550 °C. The temperature had an impact on the gas and paraffin yields. Increasing temperature up to 500 °C production of gaseous and wax products was high. Heavy olefins and aromatic compounds were found at 550 °C. C_{10} and C_{20} range of hydrocarbons were obtaine within the experimental concerning temperature 350 °C to 550 °C. The majority c gaseous products were discovered to be olefins. Within the temperature range, 350 °C t 400 °C HDPE was not fully cracked and formed solid residue. It had been reported tha the yield % of olefins was affected by HDPE/catalyst ratio. More olefins were found i wax for a high HDPE/catalyst ratio. It was presumed that 450 °C and 1.41 were th reasonable temperature and HDPE/impetus proportion individually for the thermo reactant breaking of HDPE with no strong buildup and less development of wax item (Mastral JF *et al.* 2006). Lin YH *et al.* concentrated on the breaking of HDPE ove HZSM-5 zeolite and non-zeolite impetuses (MCM-41 and SAHA). HZSM-5 was used a a co-additive and produced i-olefins and olefins. They had a conclusion that th production rate of volatile hydrocarbons was high in the case of zeolite catalysts (Lin YH *et al.* 2004).

Table 2.3 PONA distributions of liquid products at 450 °C with different catalysts

Type of pyrolysis	Total Paraffin	Individual Paraffin		Total olefin	Naphthene	Aromatic	Rest
		n-paraffin	i-paraffin				
Thermal Cracking	40.75	40.27	0.28	39.93	18.50	0.68	0.14
Zeolite Y (Powder)	5.39	0.00	5.39	79.92	7.68	7.01	0.00
Si-Al (Powder)	0.20	0.20	0.00	91.62	5.62	0.39	2.17
ZSM-5 (Powder)	1.63	1.51	0.12	16.08	23.55	58.75	0.01

Alumina (Powder)	32.57	32.57	0.00	50.19	14.99	1.14	1.11
Zeolite (Pellet)	25.10	20.68	4.42	49.28	12.05	8.43	5.14
Mordenite (Pellet)	31.07	30.89	0.18	57.07	11.51	0.13	0.22

An alteration of HDPE in vacuum gas oil to hydrocarbons was studied by Kargoz S *et al.* The breaking capacity of impetuses and the grade of fluid hydrocarbons delivered from co-handling were analyzed. It was explored that the sulfur content in the fluid hydrocarbons was impacted by various metallic impetuses. A high amount of lighter liquid hydrocarbons were produced by Co-Ac (cobalt on active carbon) catalyst in comparison with commercial hydro-cracking catalysts (DHC-8). The highest amount of sulfur-containing liquid products was obtained by HZSM-5 with smooth degradation in co-processing (Kargoz S *et al.* 2003).

Jan MR *et al.* reported the waste valorization of HDPE at concerning temperature 450 °C with $MgCO_3$ utilized as an impetus. To upgrade the interaction, waste HDPE was thermally and chemically analyzed in a reactor with various boundaries like temperature and polymer/impetus proportion. 92% of the liquid hydrocarbons were obtained with catalytic cracking at 450 °C In a single step. Waste HDPE was degraded into wax, and the wax is further cracked to produce liquid hydrocarbons. Valuable hydrocarbons were found from thermal and catalytic degradation. Diesel oil (C_{10}-C_{20}), gasoline (C_8-C_{10}), naphtha (C_7-C_{10}), or a mixture of potential hydrocarbons were found (Jan MR *et al.* 2010a).

Similar Investigation had been done with $BaCO_3$ as a catalyst and explored a comparative study of products regarding the commercial fuels by Jan MR *et al.* It was concluded that the optimum conditions for the production of liquid fuel or hydrocarbons were 450 °C temperature, 0.1 catalyst/polymer ratio, and 2 hours reaction time. GC/MS investigatic was utilized to decide the structure of the item oil tests. The oil samples had bee compared with available gasoline, diesel, and kerosene for the standardization of th products. It was observed that heavier fraction of products matched with diesel oil, light products matched with gasoline, and middle fraction matched with kerosene (Jan MR *al.* 2010b).

Jan MR *et al.* explored squander valorization of HDPE involving $CaCO_3$ as fundamental impetus. Temperatures, catalyst/polymer ratio, and reaction duration a played a role in the process optimization. Oil yield was obtained 52.33% with maximum conversion percentage (97.20%) of HDPE into useful hydrocarbons. Th ASTM distillation study was explored. The boiling point ranges of light naphtha, heav naphtha, and middle distillate were found to have fractions of 40%, 24%, and 36% respectively. GC/MS analysis was done to find out major constituents and it was reporte that hydrocarbons were in the range of C_7-C_{28}. Based on physicochemical properties was found that the liquid samples had similarities with the standard fuels like diese gasoline, and kerosene (Jan MR *et al.* 2013).

2.5 Mechanism and kinetics of thermal degradation of waste polymer

The degradation mechanisms of polymers are :

- Simple heating (Thermal degradation).
- Heating with catalyst (Thermo-catalytic cracking).
- Radiation (Photochemical cracking).
- Oxygen (Oxidative cracking).
- Heating with oxygen (Thermo-oxidative cracking)
- Chemicals (Chemical cracking)
- Heating with chemical (Thermo-chemical degradation)

The concept of these methods is common to change the molecular structure of polymer or cracking of polymers to obtain valuable hydrocarbons. Thermal decomposition an thermo-catalytic decomposition are the two most acceptable ways. 60% to 65% c polymers are polyethylene and polypropylene so these two plastics are widely investigated. Useful mixtures of hydrocarbons are obtained after degradation of polymers due to breaking of macromolecules into small fragments (Kim JS *et al.* 2003, Masuda T *et al.* 1999, Pinto F *et al.* 1999, Miskolczi N *et al.* 2004, Seddegi ZS *et al.* 2002, Uddin MA *et al.* 1997). There are two common methods for using a catalyst in the degradation process of plastics. One way is the mixing of a catalyst with polymers and then heated. Some researchers were investigated the thermal degradation of waste polymers by thermo-chemical method and found prominent results with production yields. Another one is that the molten polymers are fed on the catalytic bed (Sakata Y *et al.* 1997, Grieken RV *et al.* 2001, Jalil PA 2002, Hwang EY *et al.* 2002).

2.5.1 Reaction mechanism of plastic deformation

The usual weak connections in the structure of polymers cause the instability of macromolecules after heat treatment. Deformation of polymers after heating is a primal mechanism that supports the thermal degradation mechanism. Figures 2.2 and 2.3 depict the most likely processes during polymer heat degradation. Both the thermal and thermo-catalytic processes have the following elemental reactions:

(A) Initiation

(B) Formation of secondary radicals

(C) Recombination of radicals

(A) Initiation

In thermo-catalytic degradation, the beginning process is partly normal, and the C-C bond breakage is hemolytic. HDPE thermal degradation is initiated by a random scission mechanism, whereas catalytic degradation is initiated by the carbenium ion (Chan JH *et al.* 1997a).

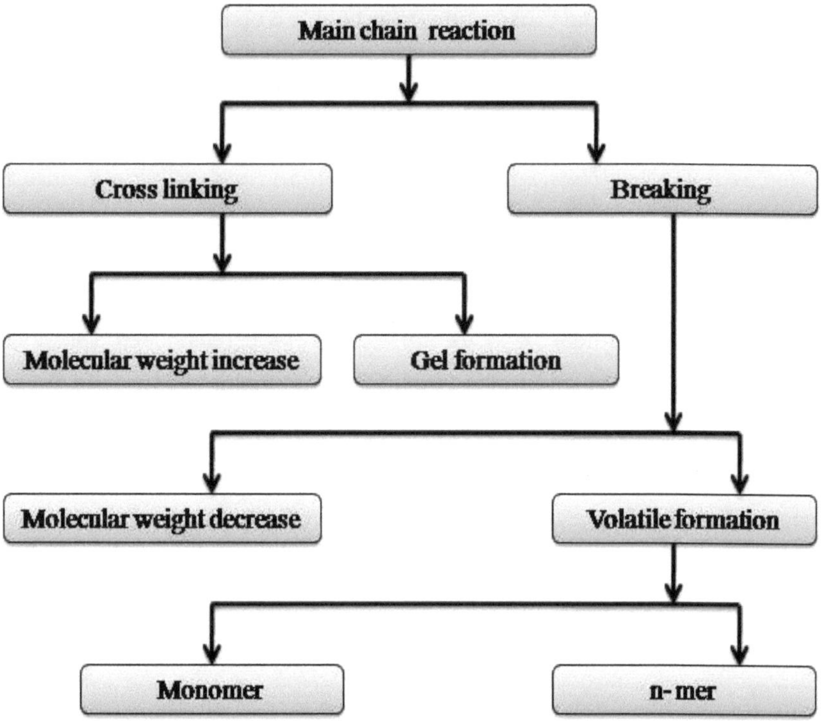

Figure 2.2 Probable main chain reactions of plastic degradation (Kumar S *et al.* 2011b

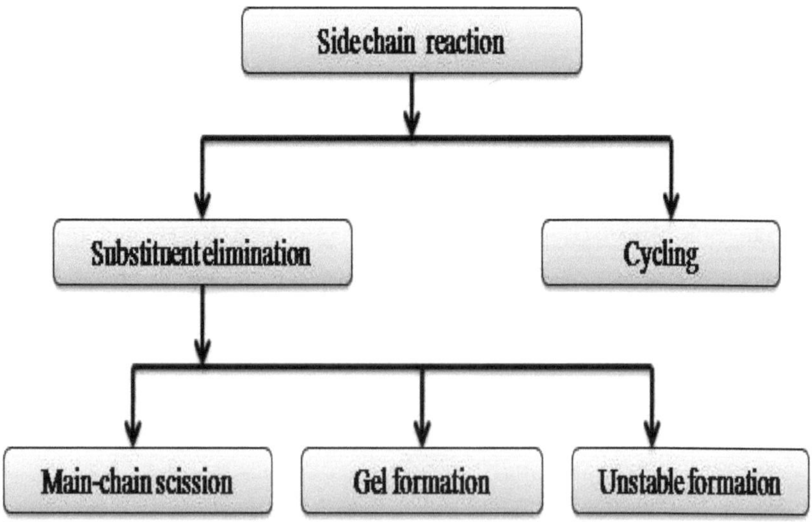

Figure 2.3 Probable side chain reactions of plastic degradation (Kumar S *et al.* 2011b

Because C-C bonds have lower bond energy than C-H bonds, their cracking consume less energy. Since the energy of C-C bond separation is more modest than that of C-H bond separation, C-C bond separation happens first. It's undeniably true that waste polymer breaking happens first for C-C bonds at lower temperatures, then for C-C, C-C, and C-H bonds at higher temperatures (Wiley J 1966). Figure 2.4 depicts the first stages of polymer heat breakdown.

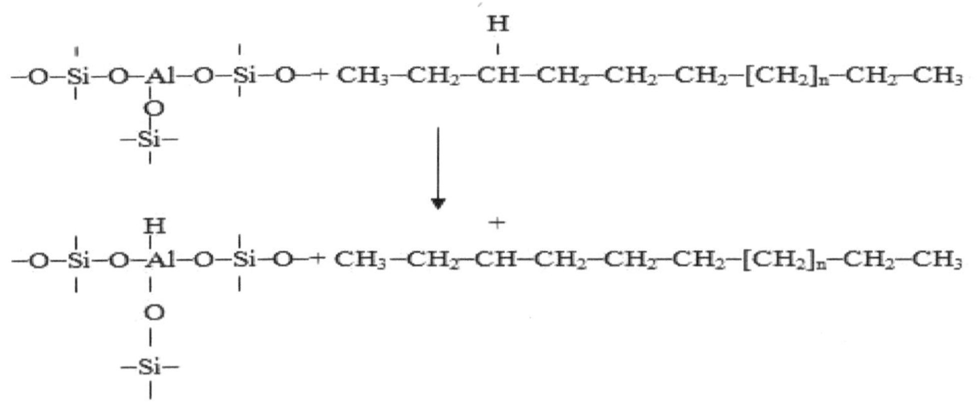

Figure 2.4. The initial step of polymer reaction (Chan JH *et al.* 1997b)

Table 2.4 shows the energy required for the elemental reactions in polymer degradation. It was observed by Moriya T *et al.* the decrement of activation energy was 50-100 kJ/mol with catalytic cracking (Moriya T *et al.* 1999).

(B) Formation of secondary radicals

Abstraction is difficult at high temperatures. At high temperatures, other processes such as intramolecular and intermolecular H-transfer, β-scission, and so on are crucial (300 °C). More amounts of lighter hydrocarbons at high temperature is caused by intermolecular H-transfer followed by β-scission. In the production of dimmers, trimers, tetramers, pentamers, and 1, 5-hydrogen transfer plays a good role (Hujuri U *et al.* 2011).

Unstable radicals and ions with uncracked macromolecules further take part in decomposition reactions. Thermal degradation of polyethylene is generally classified as "random scission" (Seger M *et al.* 1975, Seger M *et al.* 1977).

Table 2.4 Activation energies required for polymer cracking (Moriya T *et al.* 1999)

Elementary steps	Energy Required (kJ/mol)
Initiation	284-336
Diffusion of solid-state	41
Diffusion of (molten) liquid state	27
Recombination	4-10
Depropagation	21-77

Murata K *et al.* explored that there were two potential scission pathways, in view c plainly visible exploration of polyethylene, and polystyrene (Murata K *et al.* 200: Murata K *et al.* 2004).

Ueno T *et al.* reported that at high temperatures carbon number is not important for th number of scission products. They investigated the depolymerization of polyethylene a different temperature ranges. It was concluded that the β-scission becomes faste concerning high temperatures. The different temperature ranges were studied to explor the mechanism of thermal cracking. The experimental investigation was carried out a 590 °C (Ueno T *et al.* 2010).

(C) Recombination of radicals

Stabilization of primary and secondary unstable molecular fragments could be alone i different ways. According to kinetics reaction, the second-order reactions c recombination and disproportion are simple ways for the stabilization of unstabl molecular fragments. Alkenes, mono, and polynuclear molecules, or coke cyclic alkene are generally appeared in other ways of terminations like cyclization, aromatization, o polycondensation. The variable parameters to control these reactions are temperature an different physic-chemical properties of polymers have tabulated the reaction kineti parameters and reaction rates shown in Table 2.5.

- **Intermolecular hydrogen transfer reaction**

$$\cdot CH_2 \cdot CH_2 \cdot CH_2 + \cdot CH_2 \cdot CH_2 \cdot CH_2 \cdot CH_2 \cdot \xrightarrow{+} \cdot CH_2 \cdot CH_2 \cdot CH_2 + \overset{+}{CH2 \cdot CH2 \cdot CH2 \cdot CH2}$$

- **Aromatization**

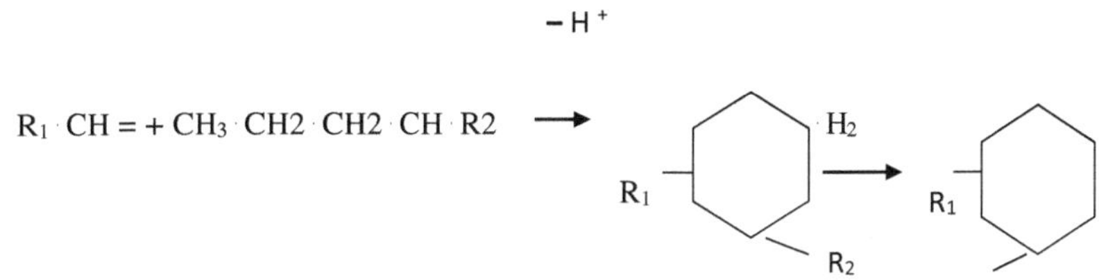

Table 2.5 Kinetic parameters for degradation of HDPE (Miranda R *et al.* 2001)

Kinetic Model	Equation	n	A (min^{-1})	Ea (kJ/mol)
HDPE $\xrightarrow{\kappa_1}$ V + R V: volatile products; R: Residue products	$\frac{d\,[HDPE]}{[HDPE]^n dt} = Ae^{[-E/RT]}$ Ea: Decomposition activation energy R: Gas constant (8.314 j/mol k T: Temperaturen: Order of reaction	0.65	1.17×10^{17}	250

- **Isomerization of the double bond**

$$- CH_2 \cdot CH_2 \cdot CH{=}CH_2 \xrightarrow{+H^+} \cdot CH_2 \cdot \overset{+}{CH_2} \cdot CH \cdot CH_3 \xrightarrow{H^+} \cdot CH_2 \cdot CH = CH \cdot CH$$

- **Intramolecular hydrogen transfer reaction**

$$-CH_2 \cdot CH_2 \cdot CH_2 \cdot CH_2 \cdot CH_2 \cdot CH_2 \cdot \overset{+}{CH_2} \rightarrow \cdot CH_2 \cdot CH_2 \cdot \overset{+}{CH_2} \cdot CH_2 \cdot CH_2 \cdot CH_2 \cdot CH_2$$

Chapter 3

EXPERIMENTATION

3.1 Introduction

This chapter delves into the procedures and methodologies that were utilized to attain the study's goals during several experimental investigations. The experimental study has been divided into the following parts. The first section deals with kaolin and sodium bentonite catalysts. The next part deals with catalyst preparation. Acid activation and base activation of the catalysts have been reported in this chapter. Different characterization methods such as FT-IR, XRD, SEM/EDAX, TGA/DTA, GC-MS, and LOI have been listed. The third section deals with the experiment setup for the production of liquid hydrocarbons. The performance and emission attributes of a solitary chamber, direct infusion diesel engine running on mixed fluid hydrocarbons are talked about in the accompanying area. In this exploratory examination, execution boundaries like brake thermal efficiency (BTE), brake specific fuel consumption (BSFC), and exhaust gas temperature (EGT) were researched, as well as outflow boundaries like carbon monoxide (CO) emanation, unburned hydrocarbons (HC) discharge, nitrogen oxides (NOx) emanation, and smoke outflow. Finally, the economic aspect of liquid hydrocarbon production from waste HDPE has been discussed. The production costs of the liquid hydrocarbons were estimated and compared with the available baseline transportation fuels.

3.2 Materials

3.2.1 Waste high-density polyethylene

Waste high-density polyethylenes from various locations were collected for this study. Following adequate washing, these plastics were dried. Plastic was crushed into smaller bits once it had dried. As represented in Figure 3.2, the gathered waste HDPE was cut into little pieces estimating around 1 cm^2 and utilized straightforwardly in this investigation. In Figure 3.1 a morphological report has been accounted for.

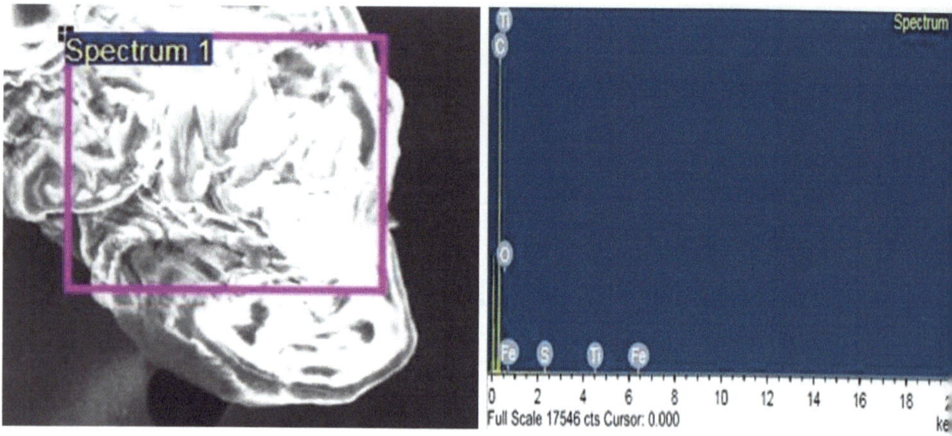

Figure 3.1 SEM image and EDAX view of HDPE sample

An OXFORD LINK ISIS energy dispersive X-ray analyzer and an LEO S430 scannin electron microscope were used for the SEM examination. A thread-like feature has bee noticed and concentrated on, as seen in Figure 3.1. The atomic percent and weig percent of the components detected in the HDPE material are shown in Table 3.1.The ar discharge method was used to cover the samples with carbon for SEM/EDAX analysi The sputter current was set to 20 mA for 59 seconds during the coal coating process. Fo SEM and EDAX investigation, 5kV and 20kV electron high tension were employe Chamber pressure was recorded as 1.0 e^{+3} mbar.

Table 3.1 Weight % and Atomic % of HDPE components

Percentage	**HDPE**				
Elements	$CaCO_3$	SiO_2	FeS_2	Ti	Fe
Weight	98.71	0.81	0.15	0.13	0.21
Atomic	99.25	0.61	0.06	0.03	0.04

3.2.2 Catalyst

Two different groups of catalysts were used in this study. One is from the kandites group of catalyst kaolin, and another one was montmorillonite groups of catalyst sodium bentonite.

3.2.2.1 Kaolin

Clay minerals, for example, are abundant in nature and may contribute to improved manufacturing. Because of their strong adsorption qualities and inexpensive cost, normally happening mud minerals like bentonite, diatomite, smectite, kaolinite, and fullers earth can be used as a substitute for enacted carbon as an adsorbent. 1:1 mud is a layered silicate structure in which oxygen molecules join an octahedral sheet of alumina to a tetrahedral sheet of silica. The development of minerals is supported by enduring occasions.

Figure 3.2 Small pieces of waste high-density polyethylene

Weathering outcomes of acidic crystalline rocks such as daccite, granite, diorite, and others, such as crude kaolin clay, are dependent on the ground surface weathering resulting from acidic crystalline rocks daccite, granite, diorite, and so on. Dirt minerals like kaolinite, bentonite, smectite, and diatomite can be used as adsorbents or as a substitution for actuated carbon. $Al_2O_7SiO_2.2H_2O$, $Al_2O_3.2SiO_2.2H_2O$, and $Al_2Si_2O_5$ are some of the chemical compositions of kaolinite (OH) 4. The kandites group of clays has a possible mineral called kaolinite. China clay, frequently known as kaolin, was given its name from the Jiangxi area express Kau-ling. It has a T-O structure, which is a nongrowing 1:1 clay silicate layered structure in which an octahedral sheet of alumina i associated with a tetrahedral sheet of silica by oxygen iotas. The O-H bunches ar professed to be practically opposite to the tetrahedral sheet, causing a wealth of hydroge bonds to shape. Kaolin is accordingly described as non-extending dirt. Kaolin has a wid scope of utilizations in numerous businesses like paper, clinical, beauty care product wine purging, synthetic, covering, and paint because of its whiteness, delicate quality and modest expense. Breaking impetus, elastic filler, wine purifier, oil adsorbent, drug plastic filler, and transporter material for insect poison and earthenware production are portion of the other significant purposes of kaolin. Kaolin or kaolinite minerals of hig purity are used to make clean white porcelain. Kaolin clay has recently been modifie and employed as a catalyst in the recycling of plastic waste. The crude kaolin has yellowish color because of the presence of iron hydroxide shades. To work on the natur of kaolin development for different modern purposes, synthetic or other chang techniques are expected to lessen or dispose of iron colors, as well as partition int constituent layers. As an additive, high-quality modified kaolin can be used to increas dispersion and improve product acceptance. Kaolin earth minerals have an atomic load o 258.071 gm/mol, a cation trade limit of 1-10 meq/100g, and a surface area of 7-30 m^2/g individually. Silica/quartz and Fe/Te-bearing minerals are plentiful in unrefined kaolin. I is, notwithstanding, promptly accessible in substance stores, with an elevated degree o pollutions like titan, iron, and potassium (Lussier RJ *et al.* 1991, Panda AK *et al.* 2011 AksharChem in Gujrat, India provided the kaolin used in this study. The kaolin i composed of 45.22% SiO_2, 38.43% Al_2O_3, 0.84% K_2O, 0.52% Fe_2O_3, 0.269% TiO_2 0.033% MgO, 0.04% CaO, 0.007% ZnO, 0.03% P_2O_5, 0.002% Cr_2O_3, 0.18% Na_2O, an 0.05% Mn_2O_3. The sodium bentonite clay and kaolin clay utilized in this experiment ar depicted in Figure 3.3.

3.2.2.2 Sodium Bentonite

The industrial application of bentonite is slightly over 132 years old; the first commercia bentonite was consigned in 1888 named "Taylorite". Mineral with high plastic clay i nature having a great number of montmorillonite was discovered in 1848 by Knight nea Fort Benton Wyoming named bentonite. The conformation of bentonite is caused by a revamping of volcanic ash or glass predominantly consisting of smectite minerals. Smectite minerals mainly consist of hectorite, beidellite, saponite, and nontronite. Two tetrahedral layers and one octahedral layer form three layers of smectite crystals. In montmorillonites, silicate layers have a negative charge as SiO_4- tetrahedrons enclosed with the M (O_5, OH) - octahedron layer. A weak negative charge is balanced by exchangeable ions. Mainly the natural form cations Ca_2^+ and Na^+ can be absorbed in the crystallite region with their hydrate shell. Montmorillonites are hydrous aluminum silicates concerned with VVV--a small amount of alkali and alkaline earth metals. Other minerals like feldspar, quartz, volcanic glass, pyrite, cristobalite, calcite, and organic matter are present in bentonite may be impurities (Bergaya F *et al.* 2006). Due to strong colloidal properties bentonite creates a viscous and gelatinous fluid with increased volume when coming into contact with simple water. Bentonites are extracted by quarrying in solid form. Initially moisture content is approximately 30% which is reduced to 15% followed by drying, crushing, and soda ash activation (not always). According to the final applications or requirements, bentonite is either milled or sieved. Today the bentonite is a well-known agent for different industries due to its special characteristics of viscosity, hydration, thixotropy, water absorption, and swelling. Due to the acceptable nature of viscosity and plasticity bentonite is used in mortars and cement (Portland). The thixotropic behavior of bentonite supports civil works such as lubricant agents in foundations, horizontal directional drilling, pipe jacking, and tunneling. The very conventional use of bentonite is the lubricating agent for cutting heads in oil and water well drilling. It is used as a mud constituent and to remove drill cuttings. Bentonite is a binding agent used in iron ore pellets manufacturing and suitable feed material in the production of direct reduction iron. Unique properties like good thermal stability, compatibility, and flowability make bentonite a good iron-industry agent in the formation of high-quality casting. After thermal treatment, it can be used as a porous ceramic carrier for various pesticides and herbicides. Oil and food industries like beer, honey, sugar, soya oil, palm oil, and canola oil use bentonite in clarifications. The environmental need is to use bentonite or the soil with low permeability as a sealing agent to protect the groundwater in the constructions and landfills. Bentonite works as the protective layer of geosynthetic clay liners. The wastewater purification has been done by bentonite due to its adsorption and absorption properties (Holt IB *et al.* 1964, Heller-Kallai L 2006 Bentonite is used in different industrial applications depending upon the speci characteristics like quantity and quality of smectites, exchangeable ions, and oth components developed by physiochemical treatments. Expanding type bentonite otherwise called sodium bentonite, and non-enlarging type bentonite, otherwise calle calcium bentonite, are two types of bentonite. The calcium type bentonite makes up mo of the bentonite found all over the planet (Asad A *et al.* 2013). Sodium bentonite is canonicalized substance with a hydrogen bond contributor count of 1 and a hydroge bond acceptor count of 13. Bentonite is an aluminum phyllosilicate that structures fro the change of volcanic debris and is essentially made out of smectite mineral montmorillonite (80% to 90% by weight). At the point when water comes into conta with montmorillonite, it grows a few times its unique volume. Bentonite's uncommo properties, like expanding and retention/adsorption, have brought its interest up in a assortment of ventures, including as a folio, paper filler, nano-clays, catalysts, sealan detoxifier, dying earth, drug items, etc. The substance construction of debris vari relying upon the kind of water it fell into. Mud minerals and earth stores have differe physical and compound properties relying upon the sort and measure of pollutio present, the idea of interchangeable cations, and the level of synthetic replacement of th smectite structure. It is challenging for at least two mud mineral developments to exi simultaneously. Besides, the compound properties of various mud tests from simila sources are not something similar (Tijen S 2010, Trauger RL 1994). Calcite, feldspa cristobalite, quartz, mica, ferrous carbonate, and pyrite are successive contaminations i bentonite (Holtzer M *et al.* 2011). Kaolin has a T-O structure and is non-extending cla though sodium bentonite has a T-O-T structure and is the growing clay. The busines sodium bentonite mud was achieved from Heilen Biopharm Pvt. Ltd, Gujrat, India. Th elements in the sodium bentonite to be specific: 0.29% K_2O, 15.34% Al_2O_3, 52.55 SiO_2, 11.92% Fe_2O_3, 1.6% TiO_2, 2.75 % MgO, 1.4% CaO, 0.06% P_2O_5, 0.01% Cr_2O and 3.22% Na_2O.

3.3 Methods

3.3.1 Catalyst preparation

Hydrochloric acid and sodium hydroxide are used respectively for acid treatment and base treatment. The NaOH (base), and hydrochloric acid used in this experiment are shown in Figure 3.4. Refluxing was done for both catalysts.

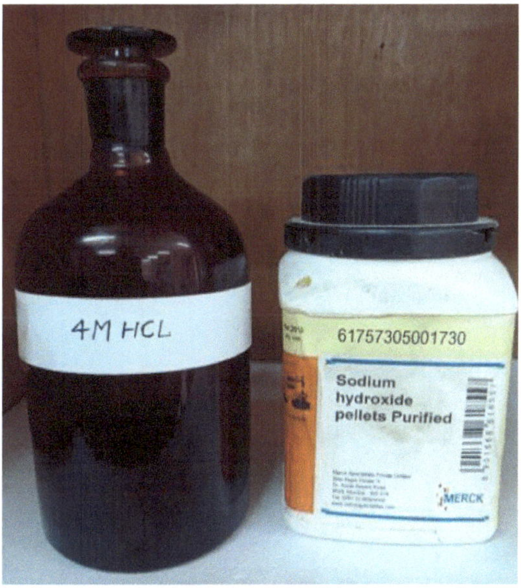

Figure 3.3 Hydrochloric acid and NaOH

3.3.1.1 Acid treatment of catalysts

60 gm of every clay sample was added to 300 ml of 4M hydrochloric corrosive arrangement and refluxed at room temperature at 105 °C for 30 minutes. For three hours, a reflux-type condenser is associated with the round-bottomed flask. Figure 3.5 (a) depicts the refluxing. After refluxing both samples, a significant quantity of cold distilled water (500-1000 ml) was added. To separate SO_4^{2-} ions from the slurry, it was centrifuged in deionized water. Calcinations were performed at 500 °C for an hour at the end of the process. K stood for untreated kaolin clay, and ATK stood for acid-treated

clay. SB and ATSB stood for sodium bentonite clay that had not been acid-treated. Th filtering process following acid-treatment has been illustrated in Figure 3.5 (b).

3.3.1.2 Base treatment of catalysts

At room temperature, probably at the cold place, 1 liter of distilled water is placed in th pyrex bottle. 160 grams of sodium hydroxide has been put slowly into the bottle and b using a magnetic stirrer mixture has been made for 4 M sodium hydroxide. 60 grams c each clay sample have been refluxed at 105 °C after putting into 300 ml of base solutio for three hours. One-hour calcinations have been done at the end of the process at 500 °C Generally calcination process was done to remove the volatile substances and to enhanc the properties of the substance. K stands for untreated kaolin clay, while BTK stands fc base-treated kaolin clay. Untreated sodium bentonite clay and base treated sodiur bentonite clay were denoted as SB and BTSB. Figure 3.6 shows the clay sample after has been calcined.

Figure 3.4 Calcination process

3.3.2 Catalyst characterization techniques

The hydrochloric corrosive actuated and sodium hydroxide base-enacted sodium bentonite and kaolin clays were explored by Fourier transformed infrared spectroscopy (FT-IR), X-ray diffraction (XRD), Scanning electron microscope/Energy Dispersive X-ray analysis (SEM/EDAX), Thermogravimetric analysis (TGA), and Differential Scanning Calorimetric analysis (DSC). To evaluate the total volatile content (loss on ignition) 50 gm of each clay sample were taken and by using a muffle furnace at the temperature of 700 °C for 1 hr the loss on ignition was calculated 12% and 9.8% for kaolin and sodium bentonite respectively.

3.3.2.1 Fourier Transform Infrared Spectroscopy (FT-IR)

FT-IR spectra have been collected by Model-SPECTRUM 100 Perkin Elmer, auto-calibrated with traceable AnC-38. The data is recorded on K-Br pellets within the range 450 cm^{-1} to 4000 cm^{-1}. The model has a horizontal attenuated total reflection and diffuses reflectance system.

3.3.2.2 X-Ray Diffraction (XRD)

The diffraction data were observed on Rigaku ULTIMA Powder X-ray Diffract meter with 2Θ measuring range -3° to 162° and minimum step size 0.0001° with rated tube voltage 20 kV to 60 kV. The geometric radius and X-ray generator of maximum output are 285 mm and 3 kilowatts respectively. XRD patterns were investigated from the range of 10° to 90°, X-ray of the wavelength of 1.540598 A° with basic accuracy ±2°. The focus size is 0.4 mm × 12 mm and parallel beam optics employ Bragg-Brentano.

3.3.2.3 Scanning Electron Microscope/ Energy Dispersive X-Ray Analysis (SEM/EDAX)

Morphology has been recorded on the LEO S430 scanning electron microscope. The energy dispersive X-ray analyzer model OXFORD LINK ISIS is attached with it. Preparation of clay samples was done by dispersing dry powder on the double-sided conductive adhesive type and scanning with the secondary electron method. Arc discharge method was used for carbon coating on the samples for SEM/EDAX.

3.3.2.4 Thermogravimetric Analysis / Differential Thermal Analysis (TGA/DTA

Model STD 2964 computerized TA machine was used for thermal analysis and α-Al_2O_3 was used as the reference material. At a constant heating rate of 10 °C/min known the weight of samples was heated in a silica crucible with a stream of nitrogen atmosphere and a flow rate of 40 ml/min from 0 °C to 900 °C.

3.3.2.5 Moisture content determination or Loss on ignition test (LOI)

50 gm of each sample was taken for the loss on the ignition test. We have the known formula to find out the loss on ignition as follows:

$$\text{Moisture content \%} = \frac{A-B}{A} \times 100$$

Where A is the weight of kaolin clay before drying and B represents the weight of the sample after drying. LOI = Loss on ignition in weight % Muffle furnace was set on before doing a test for two hours to reach the temperature at 900 °C. Kaolin clay sample and sodium bentonite samples of 50 gm were treated for 1 hr separately and weighed as 44 gm and 45.1 gm respectively by using electronic weight measuring instrument. This is done three times for separate parent clay samples. Less value of LOI for sodium bentonite shows that the acceptance performance as a catalyst used in waste plastic valorization is better than the kaolin clay. LOI values 10.98%, 12.23%, 12.65%, 11.05%, and 13.75% respectively for crude, fraction, the nonmagnetic, magnetic, and fine fraction of kaolin sample was reported by Sengupta *et al.* (Sengupta *et al.* 2008). Sodium bentonite has 9.8% LOI by weight and kaolin has 12%. 11.40%, 16.01%, 34.75%, and 18.15% are calculated values of LOI respectively of ATK, BTK, ATSB, and BTSB.

3.3.3 Gas chromatography and mass spectrometer (GC-MS) analysis

The gas chromatography and mass spectrometer were used to identify the range of trace mass hydrocarbons present in the liquid products. The extract liquid hydrocarbons have been analyzed through GC-MS model Clarus 680 GC and 600 C MS PerkinElmer, USA, at Biotech-Park, IIT Campus, Guwahati, India. A capillary column of 60 m in length are attached with the GC-MS system in presence of stationary phase compounds diphenyl (5%) and dimethylpolysiloxane (95%). Bearer gas helium was employed of flow rate 1 ml/min. Film thickness and diameter is respectively 0.25 μm, and 0.25 mm. The National Institute Standard and Technology-2008 (NIST-2008) software was employed for the interpretation of the peaks found in GC-MS.

3.3.4 Fourier Transform Infrared Spectroscopy (FT-IR) of liquid products

FT-IR equipment of model Nicolet iS10, Thermo Scientific, the USA with frequency range 400 cm^{-1} to 4000 cm^{-1} used at Biotech-park, Campus IIT Guwahati India for spectra study of liquid samples produced from the thermo-catalytic waste valorization of high-density polyethylene.

3.4 Experimental Setup and production of liquid hydrocarbons

3.4.1 Materials and items

There are following items and materials are used for the experimental setup as follows:

- Plaster of Paris
- Clay sand
- Distilled water
- Glass wool
- Heat element
- Wire
- Steel container
- Condenser
- Thermal glue/tape
- Connecting pipe
- Temperature Controller
- Thermocouple (K-type)
- Reactor (self-made)
- Switchboard
- Fire bricks

3.4.2 Required experimental setup

The experiments have been done for the optimum finding for the production of wast HDPE liquid hydrocarbons by thermo-acidic catalytic degradation, thermo-basi degradation, and simple thermal degradation (without any catalyst) processes wit thermal cracking of waste HDPE molecules. Figure 3.11 shows the layout setup for th production of waste HDPE plastic oil. 3kg of plaster of Paris and the same amount c clay sand is mixed with 2 liters of distilled water and, 200 gm glass wool dried for 1 days in sunlight for proper bonding of the materials. 45.4-gram ring diameter 5.2 mm an 720 mm length 220 volt 2500 watt silver-tone heat element wire item model numbe SYNCE008301 brand sellify is put into the bottom of the furnace. An electric furnace o 210 mm inner diameter, 240 mm outer diameter, and 30 cm height has been equippe with multi-sensor and multi-range temperature controller of model TCNS-24R Autonic with power supply range 100-240 VAC. The set value is limited to 400 °C. K-typ thermocouple temperature range 1200 °C has been attached with a temperature controlle Figures 3.7 and 3.8 depict a separate format graph of the temperature regulator's electri associations and a visual perspective on an inquiry under operating conditions. Th thermocouple is screwed with the iron-made reactor of OD 75 mm, the ID of 72 mm, an the height of 250 mm. The delivery tube of the reactor is coupled to a condenser tha leads to a measurement cylinder. The experiment is done at room temperature in the Bio diesel lab of the Mechanical Engineering Department, NERIST.

3.4.3 Production of liquid hydrocarbons from waste HDPE

Three reactors are screwed with the K-type thermocouple as shown in Figure 3.9. Eac reactor has been filled with 300 grams of waste HDPE. First is sealed with only wast HDPE without putting any catalyst. The second one is filled with acid-activated catalyst 50 grams each ATK and ATSB. The third one is filled with base-activated catalysts 5 grams each BTK and BTSB.

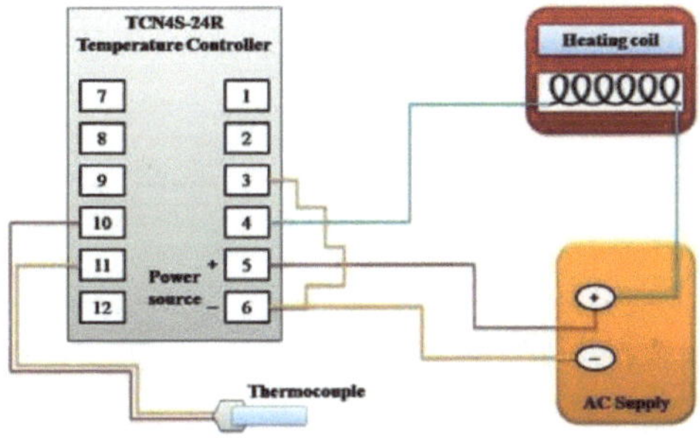

Figure 3.5 Electrical connections

The temperature has been set at 400 °C and the current supplied is about to reach the set temperature. For each run, the process continues to 90 minutes. Non-condensable flue gases, condensable oil, paraffin, and sludge have all been collected as reaction products. The gases coming out from the fuel gas chamber are allowed to pass through the liquid trap and further utilized as the heat source of the process for optimum production of liquid hydrocarbons as shown in Figure 3.11. The liquid products have been found where the heavier hydrocarbons in solid form are settled at the bottom. Figure 3.10 shows the samples collected for further analysis and characterization. AO, BO, and PO samples are representing acid-activated thermo-catalytic waste HDPE oil, base-activated thermo-catalytic waste high-density polyethylene oil, and parent oil or raw oil without catalysts respectively. The upper fluid oil of the item has been moved into a sample assortment vial the waxy item is abandoned. Also, just the buildup was left in the reactor which has been handily estimated independently for each run. The formulae used for calculating the yield and conversion percentage are given below**:** (Muthukumaran C *et al.* 2017, Yadav M *et al.* 2018).

$$\text{Total conversion (\%)} = \frac{\text{Weight of HDPE} - \text{Weight of Residue}}{\text{Weight of HDPE}} \times 100$$

$$\text{Product Yield (\%)} = \frac{\text{Weight of the product}}{\text{weight of HDPE}} \times 100$$

Finally, oil samples were prepared from the waste HDPE for a quality checkup.

Figure 3.6 Iron reactors

Figure 3.7 Liquid products obtained from waste HDPE

3.4.4 Evaluation of properties of liquid products

The Physico-chemical properties of products were evaluated as per ASTM standa specifications. The measured properties are density, kinematic viscosity, cloud poin flash point, fire point, pour point, calorific value, cetane number, and distillation. Tab 3.2 shows the test methods and acceptable ranges of fuel properties.

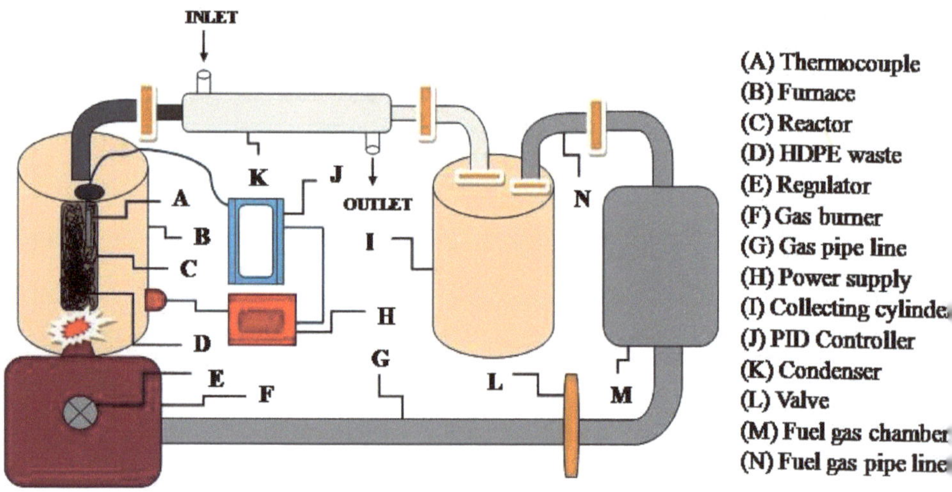

Figure 3.8 Experimental layout Diagram

Specific gravity and density measurement is done with ± 0.0005 g/ml accuracy. ±1° accuracy has been maintained during the measurements of other specifications such a fire point, flash point, pour point, and cloud point.

Table 3.2 Property test methods of fuel

Fuel property	ASTM Standard
Density @25 °C	D287
Viscosity @40 °C	D445
Calorific value	D4809
Flashpoint	D9358T
Fire point	D9358T
Specific gravity	D287

3.5 Preparation of test fuels

After confirming the properties of waste high-density polyethylene oils (AO, BO, and PO) to the ASTM specification limits, test fuels have been prepared by blending commercial diesel with obtained liquid hydrocarbons by a volume ratio of 19:1 (95% diesel + 5% AO/BO/PO). The test fuels have been prepared as AO5, BO5, and PO5. Test fuels have been stored in plastic containers for 10 days to check the miscibility of waste HDPE oils with diesel fuel. Further, some important properties have also been evaluated for different test fuels and presented in the results and discussion chapter. The prepared test fuels are as follows:

- AO5 = 95 volume % Diesel + 5 volume % Acid treated thermo-catalytic Oil
- BO5 = 95 volume % Diesel + 5 volume % Base treated thermo-catalytic Oil
- PO5 = 95 volume % Diesel + 5 volume % Parent Oil
- CD1 = Commercial Diesel (100 % volume)

3.6 Experimental investigation on diesel engine

3.6.1 Test engine setup and instrumentations

Figure 3.12 depicts a schematic representation of the experimental setup. A single-cylinder, naturally aspirated, four-stroke, direct injection, water-cooled diesel engine was used in the experiment (Kirloskar make). 'Apex Innovations Pvt. Ltd., Sangli, Maharastra, India' assembled the simple agricultural diesel engine for a research engine configuration. The determination of the test engine is given in Table 3.3. Rotameter utilized for cooling water stream rate estimation for both engine and calorimeter. Th flow rates for engine cooling and calorimeter are maintained at 150 liters per hour an 100 liters per hour, respectively, according to the instructions. EGT is measured usin two K-type thermocouples. Water inlet/outlet temperatures are measured using fou PT100 RTD-type sensors. The test fuels were injected into the combustion chamber usin a three-hole plunger type injection system (made by Bosch). A data acquisition devic was used to connect all of the signals from the sensors and transmitters to a compute (DAD). DAD takes the information and converts it into the proper format, which is the presented on the monitor. The emanations were measured using an AVL DITEST 100 exhaust gas analyzer and an AVL 437 smoke meter. Figure 3.13 shows the engin configuration (picture).

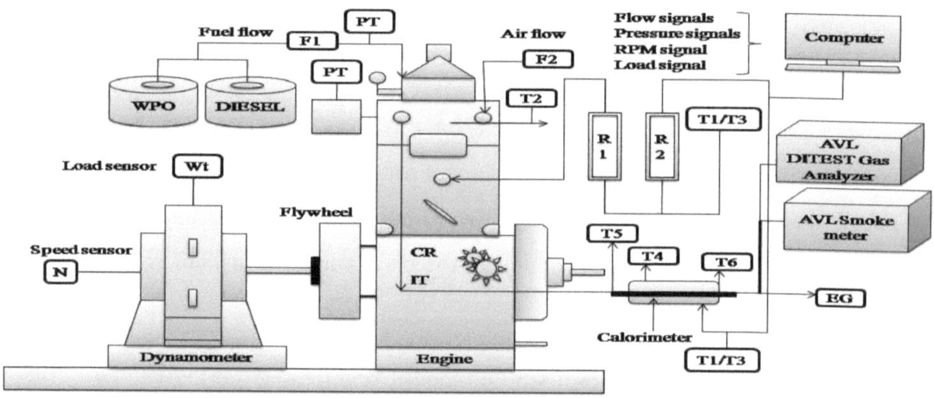

Figure 3.9 Schematic view of test engine setup

Where, T1: Water inlet temperature to the engine

T2: Water outlet temperature from the engine

T3: Water inlet temperature from the calorimeter

T4: Water outlet temperature from the calorimeter

T5: Exhaust gas to calorimeter inlet temperature

T6: Exhaust gas from calorimeter outlet temperature

F1: Fuel flow measuring sensor

F2: Airflow measuring sensor

PT: Piezo sensor

Wt: Load sensor

R1/R2: Rotameter

EG: Exhaust gas

N: RPM sensor

Technical specifications of instruments are given in the respective appendix as given below.

- Rotameter: Appendix A
- Pump: Appendix B
- Piezo sensor: Appendix C
- Eddy current dynamometer: Appendix D
- Load cell: Appendix E
- DP transmitter: Appendix F
- AVL smoke meter: Appendix G
- AVL DITEST gas analyzer: Appendix H

3.6.2 Presentation of measured test data

3.6.2.1 Performance measurement

The performance parameters are measured by collecting data from the computerized system after setting the engine at the required load level, rpm, and other operating conditions. The test fuel density and calorific value are incorporated into the software before taking any readings. BTE and EGT are mentioned as performance parameters in this research work. The brake power and brake thermal efficiencies are calculated using the following formulae:

- $$BP = \frac{2 \times \pi \times \text{Engine speed} \times \text{Engine load} \times \text{Dynamometer arm radius}}{60 \times 1000} \text{ kW}$$

Where engine speed, engine load and dynamometer arm radius are in rpm, kg-m/s^2, and m respectively.

- $$BTE = \frac{\text{Brake Power} \times 3600 \times 100}{\text{flow rate} \times \text{Calorific value}} \%$$

Where rate flow is in kg/hr and CV is in kJ/kg of the test fuel.

3.6.2.2 Emission measurement

The AVL DITEST 1000 exhaust gas analyzer is used to evaluate carbon monoxide (CO unburned hydrocarbon (HC), and nitrogen oxide (NOx) emissions from engines (NOx The smoke opacity (percent) measurement displayed in Figure 3.14 was taken with a AVL 437 smoke meter (range: 0-100 percent opacity). NOx emissions were measure using an electrochemical sensor, whereas HC and CO emissions were measured using non-dispersive infrared (NDIR) sensor. The following formulae are used for tl conversion of HC and CO emission data from ppm or volume % to g/kWh.

- CO emission in (g/kWhr) = (mf + ma) (0.345) (CO in volume %) (28/BP)
- HC emission in (g/kWhr) = (mf + ma) (34.5) (HC in ppm) (13/BP)

The dry, moister-free exhaust gas was filtered and fed through sensors for pollutic assessments. Before taking any measurements, the AVL DITEST 1000 exhaust g analyzer was given a leak check and an HC residue test. When the temperature of tl measurement chamber reached at least 70 °C, data from the smoke meter was collected.

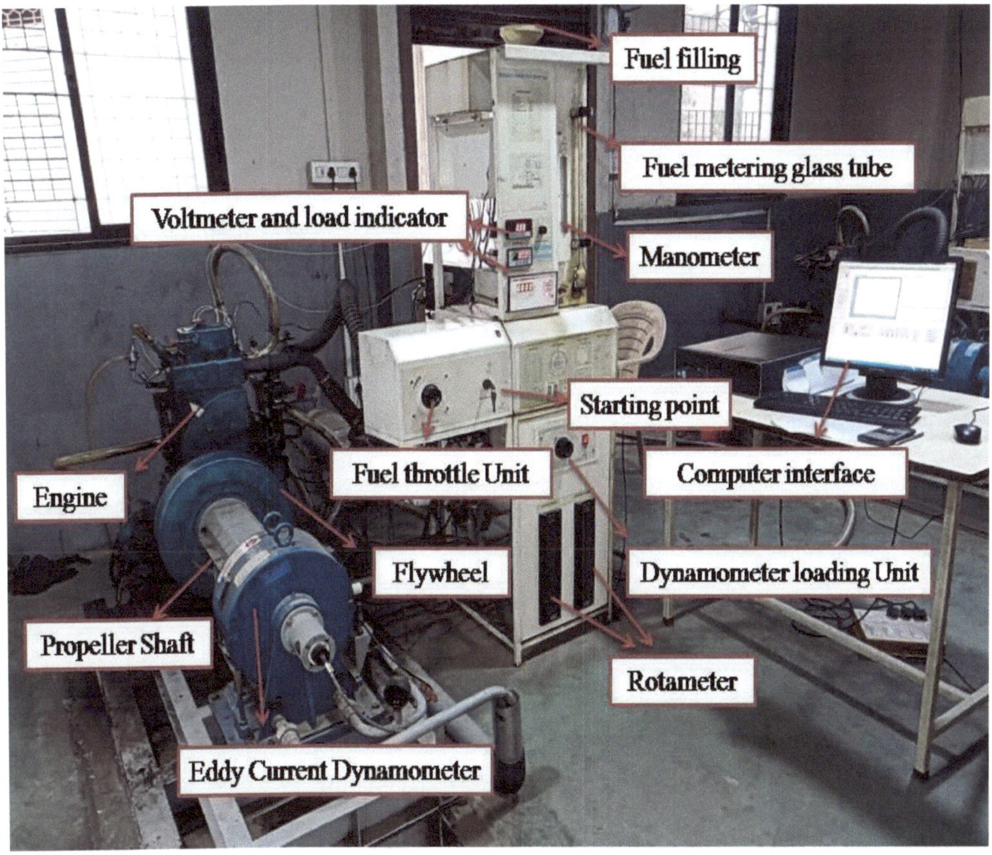

Figure 3.10 Pictorial view of experimental engine setup

3.6.3 Experimental procedure

3.6.3.1 Baseline data generation

Commercial diesel fuel is selected as baseline data generation. In the first place, th motor was turned over and heated up with diesel fuel for thirty minutes. Following th perfect activity of the motor, the calorific worth, density, dynamometer a safe distanc and different boundaries were tried for diesel in the enginesoft programming setting The engine was loaded from zero to full load in 25 percent increments by adjusting th dynamometer loading unit. 0 kg (0%), 3 kg (25%), 6 kg (half), 9 kg (75%), and 12 k (100 percent) are the proper percent load values. Following five minutes of moto running, information (from the PC show, gas analyzer, and smoke meter) were taken fo each loading, ensuring that adequate value in the noticed qualities had been laid out. A part of the repeatability check, all engine tests are run three times. The obtained result were found to be repeatable within the acceptable range. For analysis, the average of th three data was chosen and given in this study.

3.6.3.2 Blended fuel optimization

Blended test fuels such as AO5, BO5, PO5, and CD1 have been investigated as diese engine fuels in the test engine, and the results have been compared to baseline diesel fue operation for all load parameters. The corresponding calorific value and density ar placed in the software when the test fuel is changed from diesel to waste HDPE blen oils. The baseline diesel and waste HDPE blended oils were tested at conventional engin conditions, which included an 18 compression ratio, 23° CA bTDC injection timing, an a 210 bar injection pressure. The engine was run with commercial diesel fuel after th waste HDPE blended oil tests were completed to eliminate any traces of derived fue from the fuel injection system and restore the engine to its original state.

3.6.3.3 Uncertainty analysis

Any form of experimental endeavor is always accompanied by some degree o uncertainty (Moffat RJ 1988). As a result, uncertainty analysis is strongly advised whe determining the fitness value. The vulnerability of the known actual boundary R wa calculated using the Kline and Mc-Clintok approach in this investigation.

R = R (x_1, x_2, x_3, x_4 ... x_n); here connection, R is a component of the autonomous factors x_1, x_2, x_3, x_4 ... x_n. (n numbers). The weakness of the outcome R is resolved utilizing this technique, and the vulnerability of the other autonomous factors is determined too.

Table 3.3 Detail of the test engine arrangement

Engine parameter	Specification
Make and model	Kirloskar, model TV1
Details	Single cylinder, 4 strokes, DI, VCR, water-cooled diesel engine
Air induction	Natural aspirated
Bore and stroke	87.5 mm × 110 mm
Capacity	661 cc
Dynamometer arm length	185 mm
Connecting rod length	234 mm
Orifice diameter	20 mm
Compression ratio	17.5:1
Rated power output	3.5 kW @ 1500 rpm
Injection timing	23 °CA BTDC
Injection pressure	210 bar
Pressure sensor	Piezo sensor, 5000 psi range with low noisecable
Temperature sensor	PT100 (RTD) type and Thermocouple, type K
Fuel flow transmitter	DP transmitter, Yokogawa make, range 0-500mm WC
Airflow transmitter	Pressure transmitter, Wika make, range (-) 250mm WC
Ignition chamber	Celestial-sphere
Load cell	Strain gauge type load cell with digital indicator, range 0-50, kg, Sensortronics make
Dynamometer	Model AG10
Data acquisition device	NI USB-6210, 16 bit
Sensor of crank angle	1° resolution @ 5500 RPM
Inlet valve open	4.5 °CA before TDC
Inlet valve close	35.5 °CA after BDC
Exhaust valve open	35.5 °CA before BDC
Exhaust valve close	4.5 °CA after TDC

Allow U_R to be the vulnerability of the outcome R and u_1, u_2, u_3 … … u_n is the vulnerabilities in the free factors x_1, x_2, x_3, x_4 … x_n separately.

3.7 Cost of liquid hydrocarbons produced from waste HDPE

In general, we can get the waste HDPE worth Rs. 5-6 per kilogram from the nearby scra dealer. There may be a negligible transportation cost if we choose the production sit close to it. The lab-scale production cost of liquid hydrocarbons is presented in this stud For cost analysis, only the operating cost has been estimated and the capital cost is nc considered in this study. The cost of electricity consumed is also included to determin the liquid hydrocarbon production cost. Three product samples are further analyzed base on their production cost. For the optimal price, it should be noted that within th experiment the major source of electricity should be used as per need or till the formatio of gaseous products after the degradation process. Thereafter within the runnin condition, there is no need for any power sources. The cost of production of usefu hydrocarbons is about half of the available petroleum products or fuel resources lik diesel, petrol, kerosene, naphtha, etc.

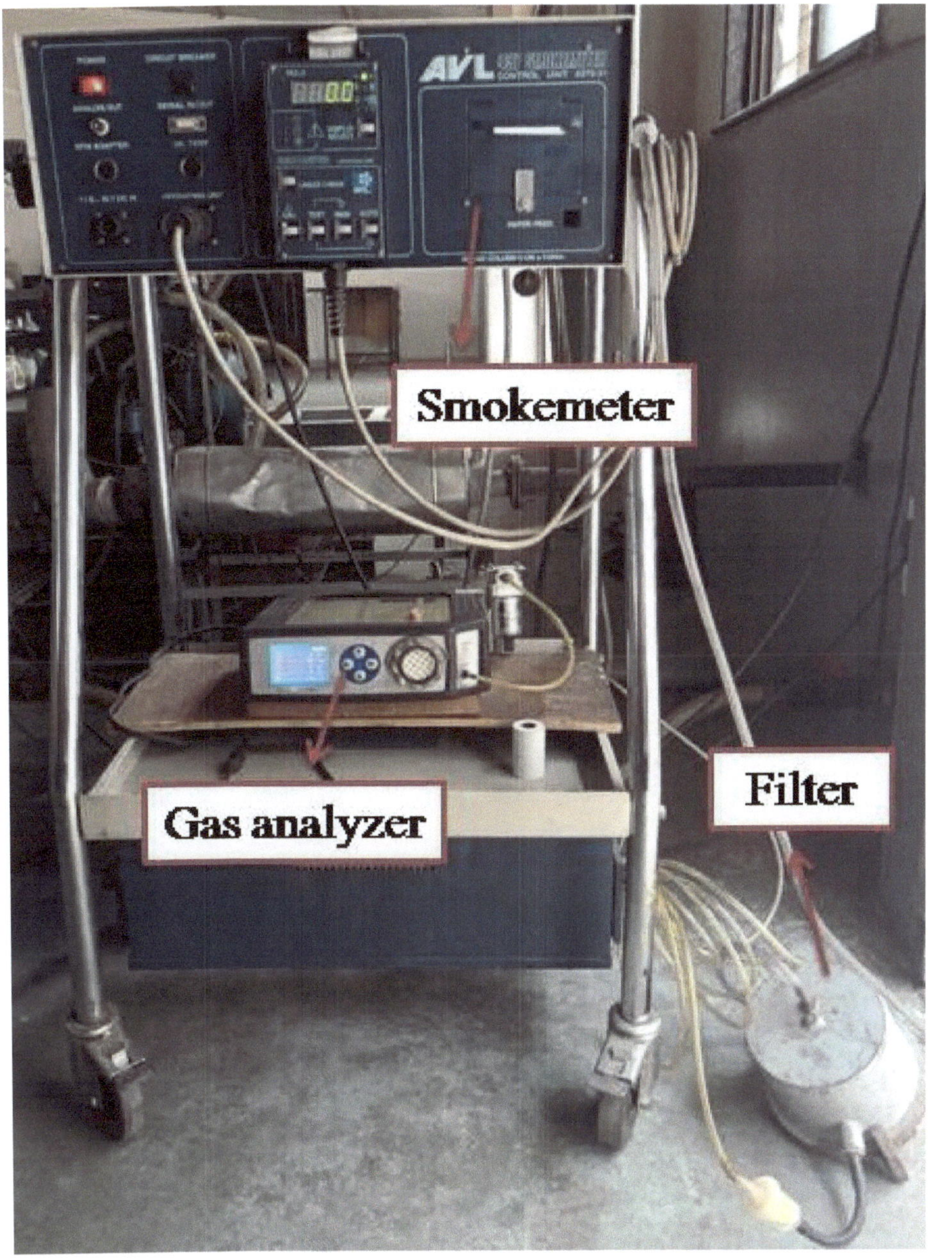

Figure 3.11 Smoke meter and gas analyzer

References

Abbas-Abadi MS, Haghighi MN, Yeganeh H. (2012) The effect of temperature, catalys different carrier gases and stirrer on the produced transportation hydrocarbons of LLDP Degradation in a stirred reactor. Journal of Analytic and Applied Pyrolysis, 95:198–204.

Achilias DS, Karayannidis GP. (2004) Water, Air & Soil Pollution: Focus, 4(5):385-396

ACRR. (2004). Good practices guide on waste plastics recycling. Brussels, Belgiun Association of Cities and Regions for Recycling.

Aguado J, Serrano DP, Miguel G. (2007) European trends in the feedstock recycling c plastic wastes. Global NEST Journal, 9:12–19.

Ali S, Garforth AA, Harris DH, Rawlence DJ, Uemichi Y. (2002) Polymer wast recycling over used catalysts. Catalysis Today, 75:247-255.

Al-Salem SM, Lettieri P, Baeyens J. (2009) Recycling and recovery routes of plasti solid waste (PSW): a review. Waste Management (New York, N.Y.), 29(10):2625-264 http:// dx.doi.org/10.1016/j.wasman.2009.06.004. PMid: 19577459.

Aroke UO, Abdulkarim A, Ogunbunka RO. (2013) Fourier transform infrare characterization of kaolin, granite, bentonite, and barite, ATBU J. Environ. Technol 6(1):42-53.

Asad A, Shantanu K, Mohammad AD, Raquibul H. (2013) Suitability of Bentonite Clay An Analytical Approach. International Journal of Earth Science, 2(3): 88-95.

Bagri R, Williams PT. (2002) Catalytic pyrolysis of polyethylene. Journal of Analytica and Applied Pyrolysis, 63:29–41.

Bari S, Yu CW, Lim TH. (2004) Effect of fuel injection timing with waste cooking oil a a fuel in a direct injection diesel engine. Proceedings of the Institution of Mechanica Engineers, Part D: Journal of Automobile Engineering. 218:93–104.

Beltrame PL, Carniti P, Audisio G, Bertini F. (1989) Catalytic Degradation of Polymers Part II-Degradation of Polyethylene. Polymer Degradation and Stability, 26:209–220.

Bergaya F, Lagaly G. (2006) General Introduction: clays, clay minerals, and clay sciences. Dev. Clay Sci., 1:1-18. http://dx.doi.org/10.1016/S1572-4352(05)01001-9.

Beyene HD. (2014) Recycling of plastic waste into fuels, a review. International Journal of Science, Technology and Society, 2(6):190-195. doi: 10.11648/j.ijsts.20140206.15.

Bhasker JS, Gopalkrishnarao P. (2010) Fourier transform infrared spectroscopic characterization of kaolinite from Assam and Meghalaya, Northeastern India. J. Modern Phy, 1: 206-210.

Brems A, Dewil R, Baeyens J, Zhang R. (2013). Gasification of plastic waste as waste-to-energy or wasteto- syngas recovery route. Natural Science, 5:695–704. doi: 10.4236/ns.2013.56086.

Chan JH, Balke ST. (1997a) The thermal degradation kinetics of polypropylene: Part II. Time temperature superposition. Polymer Degradation and Stability, 57:113–125.

Chan JH, Balke ST. (1997b) The thermal degradation kinetics of polypropylene: Part III. Thermogravimetric analyses, Polymer Degradation and Stability, 57:135–149.

Cho MH, Jung SH, Kim JS. (2010) Pyrolysis of mixed plastic wastes for the recovery of benzene, toluene and xylene (BTX) aromatics in a fluidized bed and chlorine removal by applying various additives, Energy Fuels, 24(2):1389-1395. https://doi.org/10.1021/ef901127v.

Consea JA, Font R, Marcilla A, Garcia AN. (1994) Pyrolysis of polyethylene in a fluidized bed reactor. Energy Fuels, 8:1238-1246.

Dewi R, Agusnar H. (2018) Characterization of technical kaolin using XRF, SEM, XRD, FTIR, and its potentials as industrial raw materials: SEMIRATA- First International Conference on Science and Technology, IOP Science, Medan, 5 May 2018; IOP Conference Proceedings 1116; Phys.: J. of Phy. Conference series, 2018.

Di Blasi, C. (2000). Dynamic behaviour of stratified downdraft gasifier. Chemical Engineering Science, 55(15):2931–2944.

Dudkin BN, Loukhina IV, Avvakumov EG, Isupov VP. (2004) Application mechanochemical treatment of disintegration of kaolinite with sulphuric acid. Chemic and Sustainable Development,12:327-330.

Gao Z, Amasaki I, Nakada M. (2003) A thermogravimetric study on thermal degradatio of polyethylene. Journal of Analytical and Applied Pyrolysis, 67(1):1–9.

Garforth AA, Fiddy S, Lin YH, Siakhali AG, Sharratt PN, Dwyer J. (1997) Catalyt degradation of high density poliethylene: an evaluation of mesoporous and microporo catalysts using thermal analysis. Thermochimica Acta, 294:65–69.

Garforth AA, Jesus SA, Martınez H, Akhan A. (2004) Feedstock recycling of polym wastes. Current Opinion in Solid State Material, 8:419–425.

Garforth AA, Lin YH, Sharratt PN, Dwyer J. (1998) Production of hydrocarbons b catalytic degradation of high density polyethylene in a laboratory fluidised-bed reacto Applied catalysis A: General, 169:331-342.

Georges-IVO, EE. (2005) Fourier transform infrared spectrometry and X-ray powd diffraction as complementary techniques in characterizing clay size fraction of kaolin. Appl. Sci. Environ. Manag., 9(2):43-48.

Grieken RV, Serrano DP, Aguado J, Garcya R, Rojo C. (2001) Thermal and catalyt cracking of polyethylene under mild conditions. Journal of Analytical and Applie Pyrolysis, 58–59:127–142.

Heller-Kallai L. (2006) Handbook of clay science, 1st ed.; Elsevier Science, USA, 200 pp.1000-1012.

Hernandez MR, Garcıa AR, Gomez A, Agullo J, Marcilla A. (2006) Effect of residence

Holman JP. (2011) Experimental methods for engineers, 8th ed. Tata McGraw Hill, Ne Delhi.

Holt IB, Cutler, IB, Wadsworth ME. (1964) Clays and clay minerals: 12th Nation Conference Clays and Clay Minerals.

Holtzer M, Bobrowski A, Grabowska B. (2011) Montmorillonite: A Comparison of Methods for its Determination in Foundry Bentonites. Metabk, 50(2):119-122.

Htet MT. (2010). Recycling of waste plastic in a two step thermo-catalytic reaction systems (Ph.D. thesis). http://dx.doi.org/10.4236/jmp.2010.14031.

https://www.climateaction.org/news/97-of-plastic-bottles-are-recycled-in-norway.

Hujuri U, Ghoshal AK, Gumma S. (2011) Temperature-Dependent Pyrolytic Product Evolution Profile for Polypropylene. Journal of Applied Polymer Science, 119(4):2318–2325.

Hwang EY, Kim JR, Choi JK, H. Woo HC, Park DW. (2002) Performance of acid treated natural zeolites in catalytic degradation of polypropylene. Journal of Analytical and Applied Pyrolysis, 62:351–364.

Hwang G, Kim K, Bae S, Yi S, Kumazawa H. (2001) Degradation of high density polyethylene, polypropylene and their mixtures in supercritical acetone. Korean Journal of Chemical Engineering, 18(3):396-401.

Isci S, Unlu C, Atici O, Gungor A. (2006) Rheology and structure of aqueous bentonite polyvinyl alcohol dispersion. J. Idah. Acad. Sci., 29 (5): 454-456.

Jalil PA. (2002) Investigations on polyethylene degradation into fuel oil over tungstophosphoric acid supported on MCM-41 mesoporous silica. Journal of Analytical and Applied Pyrolysis, 65:185–195.

Jan MR, Shah J, Gulab H. (2010a) Degradation of waste high-density polyethylene into fuel oil using basic catalyst. Fuel, 89:474-480.

Jan MR, Shah J, Gulab H. (2010b) Catalytic degradation of waste high-density polyethylene into fuel products using BaCO3 as a catalyst. Fuel Processing Technology, 91:1428-1437.

Jan MR, Shah J, Gulab H. (2013) Catalytic conversion of waste high-density polyethylene into useful hydrocarbons. Fuel, 105:595-602.

Karagoz S, Yanik J, Uçar S , Saglam M, Song C (2003) Catalytic and therm degradation of high-density polyethylene in vacuum gas oil over non-acidic and acidi catalysts. Applied Catalysis A: General, 242:51-62.

Kim D, Shin S, Sohn S, Choi J, Ban B. (2002) Waste plastics as supplemental fuel in th blast furnace process: improving combustion efficiencies. Journal of Hazardou Materials, 94(3):213–222.

Kim JS, Lee WY, Lee SB, Kim SB, Choi KJ. (2003) Degradation of polystyrene wast over base promoted Fe catalysts. Catalysis Today, 87:59–68.

Kline SJ, McClintock FA. (1953) Describing uncertainties in single-sample experiment 75:3-8.

Kumar S, Panda AK, Singh RK. (2011b) A review on tertiary recycling of high-densit polyethylene to fuel, Resources, Conservation and Recycling. 55(11):893-910.

Kumar S, Panda AK, Singh RK. (2013) Preparation and characterization of acid an alkali treated kaolin clay, Bulletin of Chemical Reaction Engineering and Catalysis, 8(1 61-69.

Kyrikou I, Briassoulis D. (2007) Biodegradation of agricultural plastic films: A critic review. Journal of Polymers and the Environment, 15:125–150.

Lee KH, Jeon SG, Kim KH, Noh NS, Shin DH, Park J, Seo YH, Yee JJ, Kim GT (2003b) Thermal and catalytic degradation of waste high-density polyethylene usin spent FCC catalyst. Korean Journal of Chemical Engineering, 20(4):693–697.

Lee, KH. (2006) Thermal and catalytic degradation of waste HDPE. In J. Scheirs & W Kaminsky (Eds.), Feedstock recycling and pyrolysis of waste plastics (pp. 129–160 Hoboken: Wiley.

Lin YH, Yang MH, Yeh TF, Ger MD. (2004) Catalytic degradation of high densit polyethylene over mesoporous and microporous catalysts in a fluidized-bed reacto Polymer and Degradation Stability, 86:121-128.

Lin YH, Yang MH, Yeh TF, Ger MD. (2004) Catalytic degradation of high density polyethylene over mesoporous and microporous catalysts in a fluidized-bed reactor. Polymer and Degradation Stability, 86:121-128.

Lucas N, Bienaime C, Belloy C, Queneudec M, Silvestre F, Saucedo JEN. (2008) Polymer biodegradation: mechanisms and estimation techniques – a review. Chemosphere, 73(4):429–442.

Lussier RJ. (1991) A *novel clay-based catalytic material—preparation* and properties, Journal of *Catalysis*, 129: 225–. 237. http://dx.doi.org/10.1016/0021-9517(91)90026-Z.

Mani M, Subash C, Nagarajan G. (2009a) Performance, emission and combustion characteristics of a DI diesel engine using waste plastic oil. Applied Thermal Engineering, 29:2738-2744.

Manos G, Garforth A, Dwyer J. (2000) Catalytic degradation of high-density polyethylene over different zeolitic structures. Industrial & Engineering Chemistry Research, 39:1198-1202.

Marcilla A, Beltran M, Conesa JA. (2001) Catalyst addition in polyethylene pyrolysis thermogravimetric study. Journal of Analytical and Applied Pyrolysis, 58–59: 117–126.

Mastral JF, Berrueco C, Gea M, Ceamanos J. (2006) Catalytic degradation of high density polyethylene over nanocrystalline HZSM-5 zeolite. Polymer Degradation and Stability, 91:3330-3338.

Masuda T, Kuwahara H, Mukai SR, Hashimoto K. (1999) Production of high quality gasoline from waste polyethylene derived heavy oil over Ni-REY catalyst in steam atmosphere. Chemical Engineering Science, 54:2773–2779.

McCaffrey WC, Kamal MR, Cooper DG. (1995) Thermolysis of polyethylene, Polymer Degradation and Stability, 47(1):133–139.

Miranda R, Yang J, Roy C, Vasile C. (2001) Vacuum pyrolysis of commingled plastics containing PVC I. Kinetic study. Polymer Degradation and Stability, 72:469–449.

Miskolczi N, Bartha L, Deak G, Jover B, Kallo D. (2004) Kinetic model of the chemic recycling of waste polyethylene into fuels. Process safety and environmental Protectio 82:223–29.

Miskolczi N, Bartha L. (2008) Investigation of hydrocarbon fractions form waste plast recycling by FTIR, GC, EDX-RFS and SEC techniques. Journal of Biochemical a Biophysical Methods, 70:1247-1253.

Moffat RJ. (1988) Describing the uncertainties in experimental results. Experiment Thermal and Fluid Science, 1:3-17.

Mohanraj C, Senthilkumar T, Chandrasekhar M. (2017) A review on conversic techniques of liquid fuel from waste plastic materials. International Journal of Energ Research, 41(11):1534-1552. https://doi.org/10.1002/er.3720.

Moriya T, Enomoto H. (1999) Characteristics of polyethylene cracking in supercritic water compared to thermal cracking. Polymer Degradation and Stability, 65:373–386.

Murad A, Edwin CO, Hamad SA. (2016) FT-IR and SEM studies on natural Jordani Clays: 10th International Renewable Energy Storage Conference, Riyadh, Saudi Arabi pp.15-17. Proceedings of IRES International Conference 67, 2016.

Muralidharan K, Vasudevan D. (2011) Performance, emission and combustic characteristics of a variable compression ratio engine using methyl esters of was cooking oil and diesel blends. Applied Energy, 88(11):3959–3968.

Murata K, Hirano Y, Sakata Y. (2002) Basic study on a continuous flow reactor fc thermal degradation of polymers. Journal of Analytical and Applied Pyrolysis, 65:71-90.

Murata K, Sato K, Sakata Y. (2004) Effect of pressure on thermal degradation c polyethylene. Journal of Analytical and Applied Pyrolysis, 71:569-89.

Muthukumaran C, Praniesh R, Navamani P, Swathi R, Sharmila G, Manoj KN. (201 Process optimization and kinetic modeling of biodiesel production using non-edib Madhuca indica oil. Fuel, 195:217-225.

Nagarajan G, Rao AN, Renganarayanan S. (2002) Emission and performance characteristics of neat ethanol fuelled Dl diesel engine. International Journal of Ambient Energy, 23(3):149–158.

Navarro R, Torre L, Kenny JM, Jimenez A. (2003) Thermal degradation of recycled polypropylene toughened with elastomers. Polymer Degradation and Stability, 82:279 290.

Padmaja P, Anilkumar GM, Mukundan P, Aruldhas G, Warrier KGK. Characterization of stochiometric solgel mullite by Fourier transform infrared spectroscopy. Int. J. Inorg. Mater., 3: 693-698.

Panda AK, Singh RK, Mishra DK. (2010a) Thermolysis of waste plastics to liquid fuel; A suitable method for plastic waste management and manufacture of value added products- A world prospective. Renewal and Sustainable Energy Review, 14:233–248.

Panda AK, Singh RK. (2011) Catalytic performances of kaoline and silica alumina in the thermal degradation of polypropylene. Journal of Fuel Chemistry and Technology, 39(3):198-202.

Panda AK. (2011) Studies on process optimization for production of liquid fuels from waste plastics. PhD Thesis, http://ethesis.nitrkl.ac.in/3850/.

Park JW, Kim JH, Seo G. (2002) The effect of pore shape on the catalytic performance of zeolites in the liquid-phase degradation of HDPE. Polymer Degradation and Stability, 76:495–501.

Patni N, Shah P, Agarwal S, Singhal P. (2013) Alternate strategies for conversion of waste plastic to fuels. ISRN Renewable Energy, 1-7.

Pinto F, Costa P, Gulyurtulu I, Cabrita I. (1999) Pyrolysis of plastic waste: 2. Effect of catalyst on product yield. Journal of Analytical and Applied Pyrolysis, 51:57–71.

Saikia BJ, Parthasarathy G. (2010) Fourier Transform Infrared Spectroscopy Characterization of Kaolinite from Assam and Meghalaya, Northeastern India. J. Mod. Phys., 1: 206-210.

Saikia NJ, Bharali DJ, Sengupta P, Bordolloi D, Goswamee RL, Saikia PC, Borthak PC. (2003) Characterization, beneficiation and utilization of a kaolinite clay from Assar India. Applied Clay Science, 24:93-103. http://dx.doi.org/10.1016/S016 1317(03)00151-0.

Sakata Y, Uddin MA, Muto A, Kanada Y, Koizumi K, Murata K.(1997) Catalyt degradation of polyethylene into fuel oil over mesoporous silica (KFS-16) catalys Journal of Analytical and Applied Pyrolysis, 43:15–25.

Sarker M, Rashid MM, Molla M Rahman MS. (2012) A New Technology Proposed t Recycle Waste Plastics into Hydrocarbon Fuel in USA. International Journal of Energ and Environment (IJEE), 3(5):749-760.

Seddegi ZS, Budrthumal U, Arfaj AAA, Amer AMA, Barri SAI. (2002) Catalyt cracking of polyethylene over all-silica MCM-41 molecular sieve. Applied Catalysis A General, 225:167–176.

Seeger M, Cantow HJ. (1975) Thermische spaltungsmechanismen in homo-un copolymeren aus a-olefinen. Makromolekulare Chemie-Macromolecular Chemistry an Physics, 176:1411-1425.

Seeger M, Gritter RJ. (1977) Thermal decomposition and volatilization of pol (aolefines). Journal of Polymer Science, 15:1393-1402.

Sekine Y, Fujimoto K. (2003). Catalytic degradation of PP with an Fe/activated carbo catalyst. Journal of Material Cycles and Waste Management, 5(2):107–112.

Sengupta P, Saikia P, Borthakur PC. (2008) SEM-EDX characterization of iron-ric kaolinite clay. J. Sci. Ind. Res. (India), 67: 812-818.

Seo YH, Lee KH, Shin DH. (2003) Investigation of catalytic degradation of high-densit polyethylene by hydrocarbon group type analysis. Journal of Analytical and Applie Pyrolysis, 70:383-398.

Sharratt PN, Lin YH, Garforth AA, Dwyer J. (1997) Investigation of the catalytic pyrolysis of high-density polyethylene over a HZSM-5 catalyst in a laboratory fluidizedbed reactor. Industrial & Engineering Chemistry Research, 36:5118–5124.

Silvério FO, Barbosa LCA, Piló-Veloso D. (2008) A pirólise como técnica analítica. Quimica Nova. 31(6):1543–1552.

Simic V Uhlik P. (2006) Crystallite size distribution of clay minerals from selected Serbian clay deposits. Annales Geologiquies De La Peninsula Balkanique, 67: 109-116.

Singh B, Sharma N. (2008) Mechanistic implications of plastic degradation. Polymer Degradation and Stability, 93:561-584.

Singhad RP, Tyagib VV, Allen T, Ibrahim MH, Kothari R. (2011). An overview for exploring the possibilities of energy generation for municipal solid waste (MSW) in Indian scenario. Renewable and Sustainable Energy Reviews, 15(9):4797–4808. http://dx.doi.org/10.1016/j.rser.2011.07.071.

Soloiu A, Yoshinobu Y, Masakatsu H, Kazuie N, Yasuhito M, Yasufumi A. (2000) The investigation of a new diesel produced from waste plastics ISME.

Swift G, Wiles D. (2004) Degradable polymers and plastics in landfill sites. Encyclopedia of Polymer Science and Technology, 9:40–51.

Tijen S. (2010) Purification and Modification of Bentonite and its use in Polypropylene and Linear Low-Density Polyethylene Matrix Nanocomposites. A PhD thesis submitted to the Chemical Engineering Department, Middle EastTechnical University.

Trauger RL. (1994) The Structure, Properties and Analysis of Bentonite in Geosynthetic Clay Liners. Geosynthetic Resins, Formulation and Manufacturing. Proceedings of 8th GRI Conference.

Uddin MA, Koizumi K, Murata K, Sakata Y. (1997) Thermal and catalytic degradation of structurally different types of polyethylene into fuel oil. Polymer Degradation and Stability, 56 (1):37–44.

Ueno T, Nakashima E, Takeda K. (2010) Quantitative analysis of random scission and chain-end scission in the thermal degradation of polyethylene. Polymer Degradation and

Stability, 95:1862-1869.

Venuto PB, Habib ET. (1979) Fluid catalytic cracking with zeolite catalysts, New Yor Marcel Dekker Inc.

Walendziewski J, Steininger M. (2001) Thermal and catalytic conversion of was polyolefins. Catalysis Today, 65:323–330.

Wallis MD, Bhatia SK. (2007) Thermal degradation of high density polyethylene in reactive extruder. Polymer Degradation and Stability, 92(9):1721-1729.

Wiley J. (1966) Encyclopedia of Polymer Science and Technology, New York.

Yadav M, Upadhyay S, Sharma YC. (2018) Process optimization, kinetics of productic Jatropa curcus methyl ester, and its utilization in single cylinder diesel engine. Energ Convers. Manag., 160:364-374.

Zhang GH, Zhu JF, Okuwaki A. (2007) Prospect and current status of recycling was plastics and technology for converting them into oil in China. Conservation & Recyclin 50(3):231–239.

AUTHORS

Dr. Awinash Kumar

Prof. Pradip Lingfa

Dr. Santosh Kumar Dash

Prof. Prafulla Kumar Manoharan

Thanks

www.ingramcontent.com/pod-product-compliance
Lightning Source LLC
Chambersburg PA
CBHW040743170726
48282CB00023B/126